The New Evolutionary Timetable

THE NEW EVOLUTIONARY TIMETABLE

Fossils, Genes, and the Origin of Species

Steven M. Stanley

Basic Books, Inc., Publishers

NEW YORK

Library of Congress Cataloging in Publication Data

Stanley, Steven M.
   The new evolutionary timetable.

   Includes bibliographical references and index.
1. Evolution.   I. Title.
QH366.2.S69   575   81-66101
ISBN 0-465-05013-1   (cloth)
ISBN 0-465-05014-X   (paper)

To Nell

CONTENTS

LIST OF ILLUSTRATIONS

List of Illustrations

List of Illustrations

PREFACE

THE THEORY of evolution is not just getting older, it is getting better. Like any scientific concept that has long withstood the test of time, this one has suffered setbacks but, time and again, has rebounded to become richer and stronger. What I describe in this book is evidence that evolution is not quite what nearly all of us thought it to be a decade or two ago. This evidence comes largely from the record of fossils—a record that until recently was not well scaled against absolute time. The record now reveals that species typically survive for a hundred thousand generations, or even a million or more, without evolving very much. We seem forced to conclude that most evolution takes place rapidly, when species come into being by the evolutionary divergence of small populations from parent species. After their origins, most species undergo little evolution before becoming extinct.

It is only fair to report that, while this "punctuational" view has displaced the traditional "gradualistic" view in the minds of many evolutionists, there remain dissenters. Among these are some physical anthropologists, who continue to assert that modern humans have evolved by the gradual, persistent modernization of an apelike ancestor. In chapter 7 I offer opposition to this traditional portrayal of our ancestry.

This book is not designed to build a rigorous case for the punctuational model of evolution—the goal of my more technical volume, *Macroevolution: Pattern and Process,* published in 1979. Rather, in the present book I attempt to give the interested non-specialist access to the punctuational view and its implications—implications that are by no means trivial. I also explore the history of the traditional, gradualistic view. Among the most fundamental questions here is why Charles Darwin was a gradualist. I hope that my explanations for Darwin's position will be given due consideration by historians of science, and I do not mean to be critical of Darwin here. For many reasons, he could only have been a gradualist.

The emergence of the punctuational model of evolution during the past decade has at times caused acrimonious debate. This is an exciting time in the history of evolutionary science, and those of us laboring in this complex discipline can only hope that, during the next few years, important truths will float to the top of our collective crucible without occasioning undue rancor. I do not violate this wish by attacking the biblical creationists in chapter 8 of this book. The fact is, the fundamentalist creationists are parading antiscientific views falsely under a counterfeit banner of science. The recent antievolutionary efforts of the creationists constitute a grievous insult to natural science—to astronomy, as well as to geology and biology, and even to physics and chemistry, on which the other three sciences are partly founded. It is, after all, the behavior of atoms that reveals the earth to be more than four billion years old.

I express heartfelt thanks to Léo Laporte, Ledyard Stebbins, and Jim Valentine for thoughtful and kind, but sometimes dissenting, opinions of early drafts of the first seven chapters, and to Alan Walker for the same treatment of the initial version of chapter 7. Finally, I thank Jane Isay of Basic Books for her support and encouragement.

The New Evolutionary Timetable

CHAPTER

1

Introduction

THE WORD "evolution" means unfolding, and for more than a century biologists have portrayed the evolution of life as a gradual unfolding of new living things from old, the slow molding of animals and plants into entirely different forms. It was this persistent style of change that Darwin described as "The Origin of Species."

Today the fossil record—a rich store of information that was long untapped—is forcing us to revise this conventional view of evolution. As it turns out, myriads of species have inhabited the Earth for millions of years without evolving noticeably. On the other hand, major evolutionary transitions have been wrought during episodes of rapid change, when new species have quickly budded off from old ones. In short, evolution has moved by fits and starts.

I have composed the following description in order to express the traditional view of large-scale evolution:

All around us in nature, life is ceaselessly changing. Species are modified gradually, almost imperceptibly when scaled against human history, yet in the course of millions of years the accumulation of infinitesimal steps amounts to total restructuring. Thus, during some fifty million years, what early in the Eocene Epoch had been the four-toed, fox-terrier-sized "dawn horse" was molded into the large, hoofed animal that today strains in the harness of a two-legged creature of enormous intellect. The master creature descended gradually from apelike animals, in part because changed surroundings forced these simian forebears down from the trees to a life on the savannah.

This passage would fit comfortably into many accounts of evolution published in recent decades, but its credibility fades rapidly in the light of recently recognized facts. Some of these facts pertain to the ancestry of the two-legged creature who harnessed the horse, and others relate to the ancestry of the horse itself.

Let us first look briefly at the pattern of human evolution. According to the traditional view, modern humans arose by the gradual modification of an ape-like animal during an interval of several million years. Fossil evidence now challenges this simple scenario. We now know that the human family tree has included several discrete branches, and that, at certain times, two or more humanoid species have walked the earth side by side. Even more important is the fact that species of the human family have existed almost unchanged for long stretches of geological time. *Homo erectus,* our evolutionary parent or grandparent, endured for at least a million years, and our own species, *Homo sapiens,* has undergone almost no bodily change in Europe since appearing there suddenly some forty thousand years ago. I will evaluate these aspects of human evolution and others in chapter 7 of this book, but the fundamental point is easily grasped: human evolution has apparently followed a stepwise course, with important biological change occurring as one species has branched rapidly from another.

The fossil record of horses also testifies to an episodic tempo for evolution, and this is particularly notable because for decades the record of ancient horses was heralded as the classic illustration of gradual transformation. Although this fossil record, like all others, is incomplete, so that it fails to document the full history of the horse family, one of its striking revelations is great evolutionary stability for tiny dawn horses, which, as the earliest representatives of the horse family, browsed on leaves about forty million years ago. For at least three or four million years, two species of these dawn horses roamed through woodlands of western North America. In other words, populations of these small animals replicated themselves through a million generations or so without undergoing appreciable change in form. Looking to the upper end of the horse family tree, we find the same kind of evidence. In North America, at least four species of horses survived virtually unchanged for the better part of two million years, through almost the entire recent Ice Age. In Africa during the same interval of time, four or more species of the zebra variety persisted almost without modification. All of these Ice Age species belong to the *Equus* group, which includes all living horses and zebras. The *Equus* kind of animal appears suddenly in the fossil record in North American deposits less than three million years old. This

Introduction

familiar creature evolved from an ancestor of quite different form—one that had toes flanking each of its hoofs, as well as much simpler molar teeth than the modern horse. Horses of the modern *Equus* type obviously evolved rapidly, and apparently for this reason their origin is not documented by known fossil evidence. This abrupt evolutionary birth of the modern kind of animal stands in sharp contrast to the stability of established horse species.

Thus, the new message offered by the ancient remains of humans, horses, and many other animals is that evolution has occurred episodically. Most change has taken place so rapidly and in such confined geographic areas that it is simply not documented by our imperfect fossil record. The resulting view of evolution has become known as the punctuational model, while the contrasting traditional view has been labeled the gradualistic model. The punctuational model is not incompatible with what we now know of modern life on Earth. There is good evidence that certain distinctive living species of animals have formed since the dawning of modern civilization in the Middle East.

The punctuational model might appear to represent a minor modification of the traditional scheme of evolution—an esoteric adjustment that should interest only specialized practitioners of biological science. In fact, its consequences reach much farther. The punctuational view implies, among other things, that evolution is often ineffective at perfecting the adaptations of animals and plants; that there is no real ecological balance of nature; that most large-scale evolutionary trends are not produced by the gradual reshaping of established species, but are the net result of many rapid steps of evolution, not all of which have moved in the same direction; and that sexual reproduction does not prevail in the world for the reasons that have traditionally been offered.

The punctuational model has another implication of particular interest to us, as humans. Because the human family does not consist of a solitary line of descent leading from an apelike form to our species, *Homo sapiens,* we are not as special as we would like to think. Had a perceptive being somehow passively watched human evolution for the three or four million years before our species and its immediate ancestors appeared, it could not have predicted our origin. Our anatomy and the very different anatomy of our stocky Neanderthal cousins are in many ways unpredictable from what went before. We function satisfactorily, but no better than many other configurations of bones and flesh that might have evolved in our place from the distinctive species *Homo erectus,* which was our evolutionary parent or grandparent. Our sharp chin, tall forehead, and weak brow are not the results of long-term evolutionary trends. They are apparently features unique to our species. They represent evolutionary sur-

FIGURE 1.1
The Devonian trilobite *Phacops rana*, about 300 million years old, viewed from above. The compound eyes are clearly visible, and the animal is partially enrolled, like a pill bug.

prises, albeit ones that may have been favored by natural selection at a particular time and in a particular place. The gradualistic model is compatible with such reversals, but the punctuational model predicts that they should often be very abrupt and pronounced, as indeed they seem to have been in the evolution of our own family.

We thus have strong motives to understand evolution, one of the most important intellectual concepts of the Western world, and we have every reason to sit up and take notice when it appears that fossils—pieces of inanimate rock—reveal new and surprising things about our biological origins. Fossils played only a minor role in Darwin's conception of an evolutionary process,

Introduction

and their study added little to evolutionary theory during the century that followed Darwin's revolutionary publication of 1859. Today, however, the testimony of fossils is a focal point of the debate over the nature of evolution. Fossils are vestiges of ancient life, shards of long-lost biotas. Some are skeletons— shells, bones, and teeth, for example. Others are traces of activity—tracks, trails, burrows, and even defecations. Some fossils are drab and inconspicuous, but others are spectacularly beautiful. Most are found in sand or mud, or in layered rocks formed by the hardening of such sediments with time. Fossils are quite dead, but in the eye of the sapient observer they often spring to life. In these ghosts of stone we can see shapes and movements of the past. From

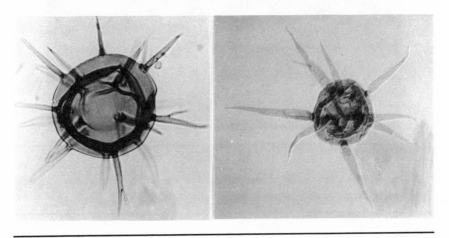

FIGURE 1.2

Algae that, more than 450 million years ago, floated in the ocean above what is now Oklahoma. These single-celled forms, and many others, have been recovered from fine-grained sedimentary deposits. (both X 1,350)

fossils, uniquely, we can read something of the history of life on Earth. Ancient patterns of behavior and of biotic interaction move within our reach.

Darwin saw things differently. He gave a dismal appraisal of what the geological record and its fossils had to offer the study of evolution. To him, the record seemed woefully deficient:

> I look at the natural geological record, as a history of the world imperfectly kept, and written in a changing dialect; of this history we possess the last volume alone, relating to only two or three countries. Of this volume, only here and there a short chapter has been preserved; and of each page, only here and there a few lines.[1]

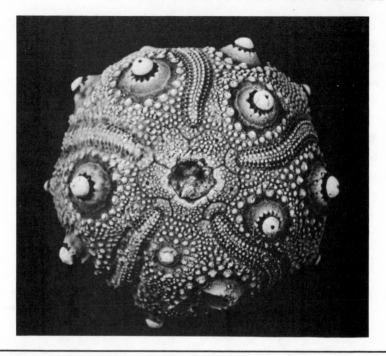

FIGURE 1.3
Top view of a fossil sea urchin of late Jurassic Age (about 150 million years old) from Germany. In life, the knobby protuberances supported spines. (×1.5)

For reasons that I will examine in the chapters that follow, Darwin's "imperfect book" metaphor was ill-chosen. The fossil record, while poor in many ways, is rich in others. It is much better understood now than in Darwin's day, and what it says to me and to many others is what I have already stated—that evolution proceeds not with the slow, gradual tempo that Darwin envisioned, but by fits and starts.

Species are basic units of life. Without having given the matter serious thought, most people know that a species is a group of animals or plants that are capable of breeding successfully with each other. To put it another way, members of one species, in general, do not breed with members of other species. A species contains a gene pool—a set of coded information that is parceled out among its members so that all develop and function in similar, but not identical, ways. Because an individual inherits half its genes from each parent, there is a constant reshuffling of genetic material from generation to generation within a species. If your mother was endowed with a gene for brown eyes from

Introduction

FIGURE 1.4
Side view of a fossil sea urchin of late Jurassic Age (about 100 million years old) from Saudi Arabia. (×1.5)

one of her parents and a gene for blue eyes from the other, it was a toss-up whether she passed one or the other gene on to you. The same holds for your inheritance from your father's side. In like fashion, chance influences your genetic legacy to your own children. The result of this kind of pairing and re-pairing of parental genes is a general genetic reshuffling within any population.

The concept of natural selection, as elucidated by Charles Darwin, is quite simple. Variability within natural populations represents the raw material. If new biological features relating to form and function are generated in the wild, as they are by experimental breeding in barnyards, then particular environmental conditions should favor certain kinds of individuals: certain kinds, among all those of a given generation, will tend to produce the largest number of offspring. Because offspring tend to resemble their parents, the character of the population will shift in the direction of the most prolific bearers of fertile offspring. Evolutionary success through parenthood actually has two facets. One is longevity: long-lived individuals produce large numbers of offspring. (Survival during the postreproductive period of life, of course, does not count.) The other facet is fecundity: a rapid rate of production of offspring can result in a large reproductive output even for an individual of average or short longevity. To have the best of both worlds in natural selection is to be blessed with unusually high fecundity for an unusually long time.

Selection of an unnatural kind is familiar to us all. This is artificial selection—the procedure employed by breeders of domestic animals and cultivated

FIGURE 1.5
Fossil fish preserved after overestimating its appetite and choking to death trying to swallow another fish. From Green River Shale of Wyoming, of Eocene Age (about fifty million years old).

plants. Artificial selection was discussed in great detail by Darwin, who invoked it to argue in favor of great power for the analogous process that he envisioned for species in nature. Artificial selection of valued strains, within what we now call animal husbandry and agronomy, is an ancient practice. Its obvious efficacy does not, however, prove the operation of powerful selection in nature. If it did, natural selection would have been recognized and widely accepted long before Darwin's time. The problem here is that artifical selection is quite severe—just as severe as its economically motivated practitioners can make it. Many young bulls are converted to steers for every one preserved for breeding. Many erstwhile race horses make a trip to the glue factory for every aging stallion preserved to stand at stud, and only a few studs are in great demand. We have no reason to assume automatically that selection pressures of comparable strength are the rule in nature. Even so, the remarkable success of artificial breeding represented a crucial analogy in Darwin's argument about what transpires in nature, and even today, artificial selection has much to tell us about evolution in the wild.

10

Introduction

FIGURE 1.6
Four-hundred-million-year-old snail from Czechoslovakia, displaying its original color pattern.
(× 3.5)

In nature, a constant environment will persistently favor particular kinds of individuals. If, however, the environment changes, evolution may track this change, provided that appropriate variability is present within the evolving population. Darwin contended that evolution tracks the environment quite closely. He believed that the variability required for the fine tuning of the organism to nature is usually present within the gene pool. I will argue that this view is no longer tenable. When well-established species survive environmental change, it is with only modest amounts of evolution. They do not maintain a tightly adjusted relationship to their surroundings.

Evolution by natural selection is really a combination of two kinds of processes, one random and the other nonrandom. The random processes, which yield all-important variability among individuals, are random in the sense of being highly accidental and unpredictable with regard to what is happening in the environment. I have already described one of these processes, the constant reshuffling of genes within a population. The other is mutation. Some mutations are spontaneous genetic changes, and others are caused haphazardly by external agents, such as cosmic rays and X-rays; most seem to represent alteration of genetic material without external causation.

Selection, then, is the nonrandom, or directional, aspect of evolution within a population. It is the trial-and-error process that determines which varieties predominate as generation follows generation. It is important to understand, however, that natural selection is not the only source of evolutionary change in nature, though it appears to be the dominant one. In small populations, chance

11

FIGURE 1.7

Fossil leaf of late Paleocene Age (almost sixty million years old). This specimen from South Dakota belongs to a species that closely resembles a species living today in Asia.

changes are not uncommon. Here, things often fail to "average out" effectively, so that accidents of survival or birthrate may sometimes overcome the force of selection. The haphazard result, known as genetic drift, must often produce changes that are neutral with respect to adaptation.

The idea now under fire is that the major effects of natural selection are brought about slowly, over myriads of generations, by the gradual transformation of well-established species. It was Darwin himself who introduced this now-traditional gradualistic view in 1859 in his classic book, *On the Origin of Species by Natural Selection,* in which the following passage appeared (p. 84):

Introduction

It may be said metaphorically that natural selection is daily and hourly scruti-
nising, throughout the world, every variation, even the slightest; rejecting that
which is bad, preserving and adding up all that is good; silently and insensibly
working, whenever and wherever opportunity offers, at the improvement of each
organic being in relation to its organic and inorganic condition of life. We see
nothing of these slow changes in progress, until the hand of time has marked the
long lapse of ages. . . .

Only after publication of the first edition of the *Origin* did Darwin introduce
the modifier "metaphorically" to the initial sentence. In an age when others
claimed a divine purpose for life, his apparent aim was to assure the reader that
such teleology had no place in his argument. His scheme was, in fact, shockingly mechanistic, not only to others but to Darwin himself.

Perhaps for its metaphorical flavor, the passage I have just quoted stands as

---

FIGURE 1.8

Top view of a fossil brachiopod shell from the Permian of Pakistan (about 250 million years
old). The spines that radiate from the lower of the animal's two valves served like a snowshoe
to support the animal in soupy mud. The upper valve served as a lid to protect the animal.
(×1.5)

one of the most eloquent in Darwin's famous book, expressing with great clarity what has become the traditional, gradualistic model of evolution. Here natural selection is portrayed as operating like an immense machine that, at the expense of speed, is geared down for power—power that vastly alters all lines of descent that survive for long intervals of time.

During the twentieth century, genetics, a field of research unknown to Darwin, has come to dominate the study of organic evolution. That gradualism has continued to prevail in evolutionary thought is illustrated by the writings of Theodosius Dobzhansky, the most eminent experimental geneticist of the mid-twentieth century. Dobzhansky, like Darwin, denied that species form rapidly:

> Instead, species arise gradually by the accumulation of gene differences, ultimately by summation of many mutational steps which may have taken place in different countries and at different times.[2]

It is a great irony that, while Darwin named his great book *On the Origin of Species,* he actually had little to say about the multiplication of species—the process we now call speciation. Rather, Darwin was preoccupied with the gradual transformation of existing species, or, in his phrase, "descent with modification." He said little about the multiplication of species except that it was a process of very slow divergence.

Most of us who adhere to the punctuational model admit that large, fully established species evolve, but we see these species as typically changing very little before becoming extinct. The lineages that they form may nonetheless sometimes change enough that the oldest and the youngest of their fossil populations cannot be accommodated comfortably within a single species. This creates what is known as the "species problem" of the fossil record. In contrast, the biologist studies only the tips of the uppermost branches of the tree of life—the only parts of the tree that are alive today. Here, while some difficulty may arise in efforts to distinguish among species, in theory every branch tip represents a separate, reproductively isolated species. Thus, in the modern world, there is no species problem comparable to the one that confronts the student of fossil life. The species problem can only be dealt with by arbitrarily dividing each dead limb of the tree of life into intergrading entities. These are called chronospecies. A chronospecies, then, is a segment of a lineage judged to encompass little enough evolution that the individuals within it can be assigned a single species name. In practice, a typical chronospecies does not exhibit a great deal more total variability, from end to end, than is found among the living populations of a similar species.

Introduction

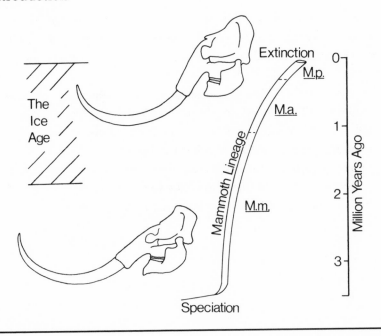

Extinction

M.p.

M.a.

The Ice Age

Mammoth Lineage

M.m.

Million Years Ago

0

1

2

3

Speciation

FIGURE 1.9

Lineage of European mammoths (genus *Mammuthus*) reconstructed from fossil data. Some paleontologists arbitrarily divide it into the three species (*Mammuthus meridionalis* (*M. m.*), *Mammuthus armeniacus* (*M. a.*), and *Mammuthus primigenius* (*M. p.*). We do not know for sure that the three species actually represent successive chronospecies of a single lineage, as shown in the diagram: possibly they overlapped in time. Certainly, the oldest species evolved very little during two and one-half million years. In any event, the head of the oldest species was not as tall as the head of the youngest species. Despite this and other minor changes, no fundamental change of form or adaptation occurred during the three and one-half million years through which the genus *Mammuthus* survived.

What, above all else, has led me to adopt the punctuational view of evolution is the simple observation that chronospecies—lineage segments that embrace little total evolutionary change—last for remarkably long intervals of geological time. In other words, it takes a very long time for a species, once established, to evolve enough to deserve a new name. On the other hand, dramatically new forms of life (species so novel as to require recognition as new genera and families, for example) spring up quite suddenly on a geological scale of time. [A genus (the singular of "genera") is the unit of classification lying one step above the species, and the family is one step above the genus; the thirty-five or so living species of the dog family (Canidae), for example, are distributed among fourteen distinctive genera, but some families of mammals contain only a single species. *Equus,* which I have discussed, is the only living genus of horses.] In order to account for the rapid origins of genera and families, we are forced to look to rapid evolutionary transitions within small populations.

In the chapters that follow, I will spell out more fully all of these ideas, which are inspired by fossil evidence. I will also review the biological evidence that supports these ideas, as well as historical reasons for their rejection by Darwin and his followers. Before examining the punctuational model of evolution and its consequences, I will, in the next three chapters, provide some historical perspective on how the gradualistic view took shape in Charles Darwin's thinking and how this view has been perpetuated to the present day.

CHAPTER

2

The Voyage Toward Evolution

TWO DAYS after the Christmas of 1831, HMS *Beagle* set sail from Devonport on a voyage around the world. Her moody and often irrational captain, Robert Fitzroy, had taken aboard an unpaid naturalist, a young man whose self-avowed goal was to gather evidence supporting his belief in the literal truth of the Bible. Little did the young naturalist suspect that scientific adventures on fossiliferous Andean scarps, in rain forests teeming with exotic forms of life, and on strangely populated volcanic islands would overturn his religious convictions. More importantly, his experiences would reshape the very foundations of biology and alter forever the human species' image of itself.

The young man, who was born February 12, 1809, the same day as Abraham Lincoln, was, of course, Charles Darwin. The notes, letters, and books that he wrote between his departure on the *Beagle* and the end of his life, half a century later, reveal much about how he came to develop what we now regard as the traditional, gradualistic view of evolution.

Among Darwin's belongings for the voyage was the first volume of Charles Lyell's revolutionary new book *Principles of Geology*.[1] Professor Henslow had supplied it as a gift which he reportedly told Darwin to read but by no means to believe. Darwin read it and believed. This and the other two volumes of

FIGURE 2.1

Ripple marks in two-billion-year-old sandstone north of Lake Huron, Ontario, Canada. These structures were produced by ocean waves.

Lyell's *Principles,* which Darwin received while in South America, became a seminal influence in his eventual conception of an evolutionary process. At a remarkably recent time in the history of our culture, Lyell's magnum opus established what is essentially the modern science of geology. With convincing, legalistic arguments (he was trained as a barrister), Lyell marshaled overwhelming evidence that the processes seen operating in nature today are the very agents that have fashioned the rocks beneath us: ripple-marks on a slab of sandstone were formed in the same way that today waves build ripples along a sandy shore. Sandstone was once sand. Shale was once mud. Volcanoes produce lava that hardens into rock; by a comparison of conformation, texture, and mineral composition, rocks formed long ago are easily identified as volcanic.

While today, Lyell's "uniformitarian" principle seems obvious (physical and chemical laws are invariant and their products change only in distribution, abundance, and configuration), in 1831 there were remnants of stern opposition. The influential and numerous catastrophists still argued for the formation of the rock record by periodic cataclysms, the last of which, by most British accounts, was Noah's flood. The view espoused by Lyell was essentially that of the Scotsman James Hutton, who in the late 1700s had set forth the concept of uniformitarianism less powerfully and in a less receptive intellectual climate.

What was especially important for Darwin's conceptual growth was that the new uniformitarian view necessarily endowed the Earth with an enormous age. Lyell viewed the planet as a huge machine, monotonously cycling materials in the manner alluded to in the biblical expression "ashes to ashes, dust to dust." Hutton, before him, could envision "no vestige of a beginning, no prospect of an end."[2] In the light of Lyell's lucid explication, the massive rock record now convinced Darwin and all other perceptive observers not only that the Earth must be extremely ancient but, that its surface changed constantly and, most importantly to Darwin, gradually. If the Earth's surface is constantly changing, why is life not forced to change in concert? If the biblical account of the Earth's formation is not literally true, what credence should be given to the biblical account of life's origins? These were the questions that began to haunt Darwin as the voyage of almost five years duration carried him through an adventure of fascinating and often startling observations.

As revealed by Darwin's *Journal of Researches,* his fortuitous geological observations, in the framework of Lyellian uniformitarianism, were nothing short of spectacular. On the night of January 19, 1835, anchored in the Bay of San Carlos, Chile, he witnessed the pyrotechnics of the volcano Osorno. He was surprised to learn afterwards that the volcano Acongagua, 480 miles to the north, had been in eruption on the same night, and he reflected on the likelihood that simultaneous events along the Andean mountain chain were causally linked. Fossil evidence seemed in accord: ". . . upraised recent [geologically young] shells along more than 2000 miles on the western coast, show in how equable and connected a manner the elevatory forces have acted."[3] On February 20, not much more than a month later, in the Chilean town of Valdivia, geological fortune again smiled upon him:

> This day has been memorable in the annals of Valdivia, for the most severe earthquake experienced by the oldest inhabitant . . . . There was no difficulty in standing upright, but the motion made me almost giddy . . . .[4]

19

The *Beagle* sailed on to Concepción, where Darwin observed the effects of the earthquake. He landed on the island of Quiriquina, where, "The ground in many parts was fissured in north and south lines . . . . Some of the fissures near the cliffs were a yard wide."[5] All around the Bay of Concepción, the land had been uplifted two or three feet. At the island of Santa Maria, thirty miles distant, there were "beds of putrid mussel-shells *still adhering to the rocks,* ten feet above high-water mark: the inhabitants had formerly dived at low-water spring-tides for these shells."[6] Darwin recognized that he was witnessing events that, on a geological scale of time, were common for the Andean region. Near Concepción, he found much older marine shell beds perhaps one thousand feet above sea level. At Valparaiso, similar beds were at a still higher altitude. The elevation of these beds must have been achieved by many small pulses of the sort he had observed. From the fact that the earthquake he had experienced was accompanied by nearby volcanic eruptions, Darwin concluded that volcanism was causally connected to uplift.

Not one geologist in a hundred has experienced earth movements as spectacular as those that Darwin chanced to witness. What he experienced were convulsive forces that were as catastrophic as any that one might encounter, yet they were not cataclysmic in the sense of being world-wide revolutions of the kind envisioned by the catastrophists, who stood in opposition to uniformitarianism. Tides had been altered locally, but the entire globe was not inundated. The Earth had quaked and parted, but the result was just a few feet of uplift. The Earth was dynamic, but its processes could easily be imagined to be cyclical. The Andes were now being uplifted, but they were also being eroded, though not as fast. Someday, if concentrated earth movements shifted away from this part of the world, erosion might subdue this particular mountain chain. What confronted Darwin in the Andes was one geological process that operated continuously but slowly (erosion) and two others that operated periodically but weakly (uplift and volcanism). Such antagonistic processes could shape the Earth over vast eons of time. This was not catastrophism in the traditional sense, but Lyellian gradualism.

The Chilean mountain chain exhibited many long-term results of such gradualism. On the Peuguenes Ridge, standing at nearly fourteen thousand feet, almost the elevation of the most lofty Rocky Mountains of North America, Darwin found seashells the approximate age of the lower Chalk in England. (These seashells were Cretaceous, or represented the last period of the Age of Dinosaurs; we now know they were likely more than a hundred million years old.) Movements of land and sea turned out to be more complex. In the

Uspallata Range, at an elevation of some seven thousand feet, Darwin came upon stumps of petrified trees standing a few feet above ground level. Nearby, above the volcanic soil on which the trees had grown, were deposits laid down beneath the sea. Evidently, the forest had first been inundated by ocean waters and then raised thousands of feet. Its silicified stumps were now being laid bare by erosion of the overlying deposits. The incessant tug-of-war between erosion on the one side and uplift and volcanism on the other could not have been more clearly portrayed to an observer of Darwin's ability.

Other geological observations, those relating to fossils, set Darwin to thinking about the history of the Earth's inhabitants. Before his paleontological efforts, few fossil remains had been unearthed in South America. Bones of the giant ground sloth, *Megatherium,* had been discovered, but little else of significance.

What was generally known about the Earth's fossil record at the time of Darwin's travels? Primary, Transition, Secondary, and Tertiary sedimentary beds were recognized, corresponding to our Precambrian or Cryptozoic (Age of Hidden Life), Paleozoic (Age of Invertebrates and Fishes), Mesozoic (Age of Dinosaurs), and Cenozoic (Age of Mammals), but in the early 1830s so meager was the knowledge of fossils that not everyone attributed to these divisions basically different forms of life. Ironically most of those who recognized a general "succession of types" were catastrophists, who until about this time were led by the Frenchman Georges Cuvier. Cuvier was not a believer in uniformitarian principles. Today he is frequently recognized as the founder of both comparative anatomy and vertebrate paleontology. With Alexandre Brogniart he studied the sequence of Tertiary sediments and fossils in the geologically youthful Paris Basin. What stood out here was an alternating sequence of marine and freshwater faunas, indicating the occurrence of large-scale relative movements of land and sea. From this regional history, Cuvier extrapolated to a world view that emphasized revolutionary events—catastrophes—which, though he saw them as regional or continental in scope rather than global, placed him at odds with Lyell and his uniformitarian followers. In 1832, shortly after the publication of Lyell's *Principles of Geology,* Cuvier was dead, but his banner had been taken up in Britain by believers in the biblical flood, who saw this as the last great catastrophe. (Cuvier himself had stood for the separation of science from religion and was thus being misused by a different group of catastrophists.)

In his brilliant anatomical work, Cuvier had advanced the theory of Correlation of Parts, according to which an organism was seen as being analogous to a machine, with each organ suited to its own purpose but correlated perfectly with all others. A hoofed animal never possessed sharp teeth for carnivory; a sharply clawed animal always did. An ambitious and persuasive academic figure, Cuvier cultivated the reputation of being able to predict the entire anatomy of any animal from a single bone (an exaggeration, but to the considerable degree that it approached the truth, a tribute to Cuvier's talent).

The principle of the Correlation of Parts implied a perfection of design related to Cuvier's belief in the immutability of species. He argued that species appeared in waves after each apocalypse. He opposed the concept of evolution, partly because of his belief in the innate perfection of species and partly because evolution implied that species disappeared not by the major extinctions he recognized but by turning into something else.

For decades, Cuvier's preeminent stature inhibited interest in evolution. In particular, he overshadowed his French colleague Jean Baptiste Lamarck, who remains famous, or infamous, for his ideas on evolution by the inheritance of structures acquired during life, often by conscious effort at improvement of adaptation. Lamarck believed that species "disappeared" by changing into new forms rather than by becoming extinct, and this was one of the ways in which he offended Cuvier. Another of Cuvier's countrymen who entertained the idea of evolution and was similarly eclipsed was Etienne Geoffroy St. Hilaire. Geoffroy emphasized similarities among organisms (in contrast to Cuvier's adaptive differences), and this led him to envision common ancestry (in contrast to Cuvier's notion that each species was a separate, perfect entity). Geoffroy's observations on monstrous features in embryos led to the suggestion that species were not anatomically strait-jacketed, but while this might have become a fruitful evolutionary avenue for further research, it was quashed by Cuvier's academic and political influence.

Ironically, from Cuvier's particular antievolutionary version of catastrophism there emerged two important concepts that helped pave the way for acceptance of the concept of evolution. One was progressionism. While Cuvier denied the modification of species, he recognized in the fossil record a trend whereby increasingly advanced organisms appeared in each wave of origination. His doctrine of progressionism had an element of truth, in that, as we now recognize, the fossil record reveals a general increase in average level of complexity and variety of organization. On the other hand, the picture is not so simple. We also know now that new kinds of simple creatures have continued to appear throughout geological time, while many have persisted from the distant past. (Similarly, Cuvier's revolutions are not without counterpart; we now recognize "mass extinctions," or brief pulses of extinction during which many major groups of animals disappeared.)

The second of Cuvier's unwitting contributions to the development of evolutionary thought related to the venerable *Scala Naturae,* or Scale of Life. This Great Chain of Being, as it has also been called, was a conceptual fabrication of Aristotle. It was supposed to link all forms of life in a graded series. A closely related concept was that of plenitude, or the fullness of nature, according to which the chain, as a perfect structure, was supposed to be without gaps. Transformation of this stationary ladder into a kind of escalator of evolution required no great act of imagination. In fact, in the seventeenth century Leibnitz had proposed the Principle of Continuity—that species did not exist as fixed types (as they had been viewed according to the Platonic ideal) but graded into one another. Leibnitz even entertained the possibility that some intergradation resulted from evolution. Cuvier, however, typified biologists and paleontologists of the early nineteenth century in retaining the idea that species were immutable.

On the other hand, Cuvier recognized four large natural groupings, or *embranchements,* of animals, which are partly retained in the classification of today (one being the Vertebrata, or animals with backbones). In this way, Cuvier unwittingly opened the way to the concept of evolutionary divergence. By fracturing the traditional scale into four pieces, he encouraged future thinkers to contemplate the branching of one piece from another, in a simple pattern of the kind that we now recognize metaphorically as the tree of life.

What were the biological views of Lyell, the man whose *Principles of Geology* educated Darwin aboard the *Beagle?* It was a critical flaw of Lyell's world view that while his essentially correct uniformitarian picture of geology entailed vast stretches of time, he opposed progressionism, or any major temporal change in forms of life. Corresponding to his dynamic cycling of Earth materials, he envisioned a balanced, nearly steady-state biotic condition. Lyell's biological balance of nature was in part an outgrowth of the idea of plenitude. Like Cuvier, Lyell saw every species as being both perfectly fitted to a particular role and immutable. Extinction occasionally overcame a species—Lyell knew the fossil record too well to believe otherwise—but this happened because of locally changed physical conditions, and the role of lost species was filled by the expansion of other species or by the appearance of altogether new, but similar, ones. In such a system, plenitude was not violated by change. Remarkably, Lyell remained mute on the origin of species. Perhaps his mechanistic view of the Earth's surface caused him discomfort with divine intervention. Certainly

FIGURE 2.2

Ichthyosaur with the outline of its skin preserved, from the Lower Jurassic of Germany (about 190 million years old). Ichthyosaurs—swimming reptiles shaped like dolphins—were among the first dramatically well-preserved vertebrate animals to be found in the fossil record of extinct marine life.

progressionism, and possibly creationism, offended him because they were basic tenets of the British catastrophists, the illegitimate descendants of Cuvier against whom Lyell stood in staunch opposition.

Whatever his protected views on creation may have been, Lyell's opposition to progressionism or any large-scale directional biotic change meant that a reading of his work provided Darwin, on board the *Beagle,* with no motivation toward an evolutionary mode of thought.

In the 1830s, the fossil record was poorly enough known for Lyell to argue with some cogency against the reality of major temporal trends. That Darwin tended to accept the reality of major trends in the face of Lyell's opposition is indicated by his employment of fossils to compare the ages of South American sedimentary deposits to ages of deposits studied in Britain. Possibly he found it difficult to abandon the belief, acquired during his Cambridge days, that the fossil record displays a succession of life forms.

In time, Lyell painted himself into a corner with his obstinate denial of the fossil record's increasingly evident general pattern of temporal biotic change. But in the 1830s, even for the vertebrates, which were the focus of progressionistic debate, the fossil evidence was only suggestive of large-scale change. The record of vertebrates had begun to display itself in the haphazard manner that typifies anything dependent on revelation by chance discovery. The first ancient "sea monster"—an ichthyosaur (a dolphin-shaped reptile)—had been unearthed at Lyme Regis in 1812, but not until 1820 was it recognized as being something other than a crocodile. In 1823, a long-necked plesiosaur was uncovered in the same area, and in 1824 the first dinosaur, the duck-billed *Iguanodon*, was disinterred elsewhere in southern England. Even the broadest outlines of the history of life were slow to take shape. Spectacular dinosaur finds, for example, awaited the opening of the American West. Some discoveries were confusing. The recognition in 1828 of an opposumlike mammal near Oxford in the Stonesfield Slate, a deposit now recognized as representing the middle of the Age of Dinosaurs, contributed to Lyell's argument: the contemporaneity of mammals with what seemed to be early reptiles militated against progressionism. We now know that even among vertebrate animals strict progressionism is an oversimplification. Some mammals were indeed thriving while the dinosaurs ruled the earth, but they were small, primitive forms. The oppossum, a member of a family that appeared during the Age of Dinosaurs, is a living fossil. Advanced mammals appeared only after the dinosaurs were gone.

Lyell's arguments relied heavily on his belief that the record of fossils is very poor. After the return of the *Beagle,* Lyell became one of Darwin's closest scientific confidantes. As we shall see, Lyell's long-standing condemnation of the fossil record became Darwin's unfounded apology for his gradualistic view of evolution and, I will argue, a major source of error in his thinking. When Darwin finally published his ideas on evolution, his arguments were based largely on information supplied by living organisms rather than by fossils. Lyell was eventually swayed by the cogency of the biological arguments and became a full believer in large-scale change and a partial believer in evolution. There was a great irony here. Lyell, the expert on the rock record, which displays quite clearly trends of large-scale biotic change, finally came to accept the fact of this change not from the record itself but from Darwin's biological evidence pointing indirectly to more limited degrees of change.

To Darwin, on his voyage, fossils had great significance at least in confirming the fact of extinction and provoking questions about this destructive process. During the eighteenth century, the reality of extinction had been questioned. It was not until 1786 that Cuvier provided convincing evidence that species had actually been lost from the Earth. Many fossil invertebrate species found in marine sediments lacked known living representatives, yet deep and distant modern seas remained unexplored. Possibly somewhere the curious fossil forms still flourished. Cuvier cogently singled out large land animals for a test. He showed that the fossil mammoths of Eurasia were not of the same species as the modern Indian or African elephants, and it was generally agreed

FIGURE 2.3

Charles R. Knight's reconstruction of the woolly mammoth, which lived in Europe during the Ice Age, just a few thousand years ago. The obvious disappearance of the mammoth from the Earth represented the first example of the extinction of a species to be recognized by natural scientists.

that animals as large as mammoths could not have been overlooked in the modern world. Extinction was real. Still, the nature and extent of extinction were little known at the time of the *Beagle's* voyage. Darwin's notebook of 1837 reveals that at this time he, like Lyell, retained the concept of plenitude, and it was in this context that he began to contemplate his own observations on extinct forms of life. He assumed the world to be brim-full of life, with species that disappeared being replaced by others that were newly formed or that existed before but now expanded their niches. Even so, Darwin's curiosity was piqued. Why had some species died out while others survived? His *Journal of Researches* was written in 1842, after his conception of natural selection, so we must beware of unconscious distortions in his retrospective account of ideas entertained during the voyage. In any event, he wrote of southern Patagonia, where fossil horses could be found, but where no modern horses lived until the Spaniards introduced them.

> What shall we say of the [New World] extinction of the horse? Did those plains fail of pasture, which have since been overrun by thousands and hundreds of thousands of the descendants of the stock introduced by the Spaniards? Have the subsequently introduced species consumed the food of the great antecedent races?[7]

The doctrine of plenitude is evident in this passage, but so is the question of numbers and food that, as we shall see, Darwin's reading of T. R. Malthus's ideas on population growth turned into the doctrine of natural selection.

Apart from providing evidence of extinction, the biological facts that Darwin assembled in his travels and afterward may be divided into two categories with respect to his formulation of a theory of evolution: evidence bearing on the simple reality of evolution and evidence bearing on the evolutionary process that he envisioned (natural selection). It is not widely appreciated that the experience of the voyage led Darwin toward the idea of evolution more than toward the idea of natural selection. Certainly Darwin's travels revealed variability among organisms—the raw material of natural selection—and he also saw elements of the biotic struggle that he later invoked as a source of selection. His observations here were of a general nature, however, and aboard the *Beagle* he could hardly have been seeking the mechanism behind a phenomenon in which he as yet did not believe. The influence of the voyage formed part of a logical sequence: first came recognition of the phenomenon, later an explanation.

The glaring biological evidence that, against Darwin's will, impelled him away from fundamentalist creationism and toward an acceptance of some form of evolution resulted largely from his observations of the geography of life. To appreciate the power of his geographical observations, one must temporarily cast aside all modern notions of evolution and trade places with Darwin—a theologically orthodox young man bent upon gathering earthly documentation for Divine Creation. In fact, what soon began to appear on the voyage were indications of something quite different from creation—evidence of apparent historical relationships between species.

Chapters Eleven and Twelve of the first edition of Darwin's *On the Origin of Species* reveal the logic of his conversion. Perhaps the most basic conundrum he was forced to contemplate was the apparent irrationality of creation. A creative divinity would have been expected to distribute its productions widely, wherever conditions allowed them to flourish. Darwin, however, noted that suitability of physical conditions alone could not explain biotic distribution (p. 346):

> There is hardly a climate or condition in the Old World which cannot be paralleled in the New—at least as closely as the same species generally require. . . . Not withstanding this parallelism . . . how different are their living productions!

The demonstration that a species had multiple origins would rule out any evolutionary mechanism, he reasoned, because whatever process was involved, nature should never yield exactly the same form twice. Nonetheless, no evidence of multiple origins appeared. In writing the *Origin,* more than two decades after the return of the *Beagle,* Darwin dealt with apparent exceptions. He pointed out, for example, that species represented by populations on widely separated alpine mountaintops had clearly been dispersed during the great glacial episode that Louis Agassiz had elucidated in the 1830s. Such populations of a single species were remnants of what had once been a continuous distribution. They were not violations of Darwin's "single place of origin" rule.

Darwin also made the more subtle observation that species having broad geographic distributions are generally species with exceptional powers of dispersal: they have achieved their broad ranges by spreading, presumably from some point of origin, not by having been scattered about by some arbitrary external agency operating without dependence on the species' inherent abilities to disperse. For example, Darwin observed that animals like freshwater insects and mollusks were wide ranging (those of Brazil resembled those of Britain), but after the voyage, he showed that certain juvenile mollusks could survive for many hours out of water on a duck's feet. Similar vectors of dispersal could easily be contemplated for wide-ranging insects. Darwin eventually concluded (and it is still generally believed) that freshwater animals have evolved unusual mechanisms of dispersal because of the value of these traits in a world that constantly changes: ponds and streams frequently dry up.

While the biotas of the Earth's several continents had little in common, Darwin noted remarkable similarities within each continent. Thus, South America harbored two species of large, flightless birds, but these were unrelated to the ostrich of Africa or the emu of Australia. Instead, they were both rheas. Why was neither an ostrich or an emu? Why were both of the same, unique biological group? Why did many of the remarkable flying birds of South American rainforests also fall into distinct groupings peculiar to the continent? Why were the several large South American rodents constructed on a single body plan? These hints of common ancestry were everywhere.

The fossil record of South America added a temporal dimension to the evidence of living faunas. Darwin was thrilled to unearth several large fossil mammal skeletons in Patagonia. Among other distinctive forms, he discovered an enormous armadillo. Interestingly, the living armadillo, a similar but much smaller species, also inhabited South America. Even more curious among the extinct animals was the giant ground sloth, which resembled the living arboreal sloth, also of the South American continent, but was many times larger and obviously could not climb trees. All of these animals are, in fact, united in a group of mammals named the Edentata ("toothless ones"). Other large fossil

FIGURE 2.4
Rhea darwinii, a rare species that Darwin learned about from the gauchos in Patagonia. John Gould, the ornithologist and artist, supplied this illustration and named the animal after Darwin.

mammals bearing obvious resemblance to living South American forms also turned up in Darwin's explorations. Why would Divine Creation adhere to a single ground plan for denizens of one part of the Earth, especially when habits changed radically through time, as in the case of the sloths? Fossil seashells belonging to living species occurred close by some of the extinct mammalian behemoths, showing that the latter lived and vanished from the Earth recently in geologic time. That giant mammals seemed preferentially to have become

FIGURE 2.5

The extinct giant ground sloth, *Megatherium,* and the giant armadillo, *Glyptodon,* Ice Age mammals whose remains Darwin studied in South America.

extinct bemused Darwin, and this kind of puzzlement was to become one element in his arrival at the concept of natural selection. More immediately, however, he was struck by additional aspects of the continuity of structure within a given area.

The peculiarities of individual continents were just as conspicuous as the evident effects of more localized barriers. Why did faunas differ on opposite sides of large rivers, mountain chains, and deserts? The distinct nature of marine life on the Atlantic and Pacific sides of the Isthmus of Panama was especially curious. Would not a divinity introduce many of the same species to both sides of such a narrow neck of land?

Probably the geographic evidence most compelling to Darwin was provided by islands. In the *Origin,* he described the apparently total absence from oceanic islands of many large groups of terrestrial organisms. He found no evidence that a single species of frog, toad, or newt had ever been native to an oceanic island. The same seemed true for land-dwelling mammals, at least with respect to islands situated more than three thousand miles from a large continent or continental island. Was it sheer accident that the only mammals found on such islands were airborne bats? Darwin observed in the *Origin* (p. 390) that, in general, insular biotas were glaringly depauperate until supplemented by human culture:

> He who admits the doctrine of the creation of each separate species, will have to admit, that a sufficient number of the best adapted plants and animals have not been created on oceanic islands; for man has unintentionally stocked them from various sources far more fully and perfectly than has nature.

The voyage of the *Beagle* provided more subtle evidence. The Galápagos Islands, between five and six hundred miles offshore from Equador, have come to symbolize Darwin's work aboard the *Beagle*. Here, Darwin observed in his *Journal of Researches* (p. 378), "both in space and time, we seem to be brought somewhat near to that great fact—that mystery of mysteries—the first appearance of new beings on this earth."

The fauna and flora of the Galápagos, which Darwin studied for more than a month, are so novel as to be spectacular. One characteristic of the Galápagos biota that Darwin viewed as evidence against creation was endemism, or uniqueness. In fact, a large percentage of Galápagos species are endemic to that group of islands. The initial question was why species should be specially created for particular archipelagos. The Galápagos bird fauna became evidence for Darwin's contention that oceanic islands have instead been populated naturally by the geographic spreading and differentiation of life. In particular, the least endemic Galápagos birds were those that would have been expected to arrive in the islands by normal dispersal (pp. 390-391):

> Nearly every land bird [of which there were twenty-six], but only two out of the eleven marine birds, are peculiar; and it is obvious that marine birds could arrive at these islands more easily than land birds.

Details of distribution offered even more remarkable patterns. It is no accident that the Galápagos Islands derive their name from the Spanish word for tortoise. Among the most famous of the unique inhabitants of the islands are the giant land tortoises, some of which Darwin estimated to exceed five hundred pounds in weight and one of which he clocked at less than one-quarter of a mile per hour. What struck him as curious was that people of the Galápagos could identify the particular island of origin of any tortoise they were shown. Each island had its own race of the large reptiles, although all races were assigned to a single species. Again, the importance of geographic barriers loomed large. Why would a creator populate the islands with so many varieties and why only one per island? Was it not more logical to postulate some natural

FIGURE 2.6
The giant tortoises that gave the Galápagos Islands their name.

mode of divergence of the separate populations, perhaps related to their unique conditions of life?

The Galápagos finches, though less spectacular as individual animals, were collectively even more remarkable than the tortoises. They comprised half of the twenty-six species of Galápagos birds. The thirteen species of finches were all endemic and shared many structural characters, but differed markedly from each other in shape of beak. The beak of an insect-eater was small. That of a nut cracker was heavy and robust; and one species instinctively employed a cactus spine to probe for grubs. Taken together, the beaks of the thirteen species formed a graded series ranging from a parrotlike structure to one that was small and thin. The adaptations to particular modes of life were in many cases obvious, and Darwin wrote in his *Journal*, published after his conversion to evolution by natural selection:

FIGURE 2.7
Galápagos finches with heavy bills, like those of grosbeaks.

Seeing this gradation and diversity of structure in one small, intimately related group of birds, one might really fancy that from an original paucity of birds in this archipelago, one species had been taken and modified for different ends.[8]

The Galápagos finches illustrated another aspect of biotic geography that in the *Origin* (p. 398) formed one of Darwin's most compelling arguments for species-to-species descent. Most of the endemic species of Galápagos birds, including the finches, betrayed close similarities to South American species "in every character, in their habits, gestures and tones of voice." Darwin continued: "Here almost every product of the land and water bears the unmistakable stamp of the American continent." From his observations on the voyage, he generalized further, claiming that his conclusions held for islands near continents throughout the world. From this relationship emerged the question of why a creating divinity would thus constitute island faunas. The Galápagos are volcanic islands and, owing to their relatively rugged surface and persistence above sea level, Darwin correctly judged them to be relatively youthful, in a geological sense. The implication, which Darwin had fully accepted within a year or so of his return to England, was difficult to resist. Species endemic to newly formed islands were derived from nearby land masses by some natural agency. Only in this manner was it possible to explain why the young insular faunas resembled nearby continental faunas in general, but were unique in detail. The islands had been colonized by very small groups of refugees whose descendants had somehow undergone modification within new surroundings so that they now represented distinct, but still similar species. If this was the normal course of events for islands, would not similar transformation be the rule for life in general? That the Galápagos observations were pivotal to Darwin's conversion to evolution, which was by no means instantaneous, is indicated by a passage inscribed in his ornithological notebooks at the time of the *Beagle's* visit to the islands. With particular reference to his previous notes on the distinctiveness of the tortoises of individual islands, he wrote, "If there is the slightest foundation for these remarks, the Zoology of Archipelagoes will be well worth examining; for such facts would undermine the stability of species."[9] The implications were staggering.

CHAPTER

3

The *Origin* and Its Very Slow Process

THE *Origin of Species,* published in 1859, twenty-three years after the return of the *Beagle,* was to many readers disarmingly powerful despite its unorthodox message. Darwin positioned the geographic arguments, described in the preceding chapter, near the end of the *Origin of Species.* These arguments converged on the idea that natural groups of organisms have descended from others in certain areas: when a distinctive new species has developed by descent with modification, it has often been ramified into additional species and they, in turn, have done likewise. Each portion of the tree of life has grown in this manner. The overall pattern of descent in nature is three dimensional, with the appearance of new branches tending to expand the two-dimensional geographic region occupied. The geological time required for diversification contributes the third dimension. As Darwin put it in the *Origin* (p. 130), employing the now familiar botanical metaphor:

> As buds give rise by growth to fresh buds and these, if vigorous, branch out and overtop on all sides many a feeble branch, so by generation I believe it has been with the great Tree of Life, which fills with its dead and broken branches the crust of the earth, and covers the surface with its ever branching and beautiful ramifications.

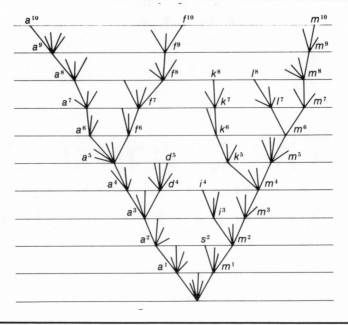

FIGURE 3.1
The tree of life published by Darwin in the *Origin* (1859, p. 117). The tree depicts a gradualistic pattern of evolution. Each fanlike pattern represents the slow evolutionary divergence
of populations. Darwin believed that new species, and eventually new genera and families,
formed by this kind of slow divergence.

While we might also metaphorically suggest that the concept of evolution
transformed the ladder of life into an escalator, the change of pattern was
actually much more drastic. As we have seen, the linear Chain of Being, in
which contiguous links had been supposed to be closely similar according to
the perfect wholeness of the divine plan, had been broken by Cuvier into four
segments. Darwin further rendered it into many connected branches. Cuvier's
four *embranchements* transgressed time, and yet they were not continuous; each
combined several waves of creation. In contrast, each branch of Darwin's tree
represented a continuum: it grew. The evolutionary model of diversification
had of course been suggested formally by predecessors of Darwin, but without
the mechanism of natural selection. Part of the appeal of the tree metaphor had
always been that it made the classification of life somewhat natural rather than
totally artificial. Then, as now, classification went much further than the simple
recognition of species. Species are grouped into genera, genera into families,
families into orders, orders into classes, and classes into phyla. Because there

can be several species in a genus, several genera in a family, and so forth, the structure of classification is inherently hierarchical. Groups of varying ranks are nested, as organizations are nested within an industrial conglomerate. A conglomerate might consist of several corporations, each of which comprises several divisions. We now see why the pre-Darwinian tradition of classifying organisms hierarchically came into being. It mirrored the actual configuration of the tree, with large branches and their ramifications representing classes, smaller divisions representing orders, and so on. Species represent small twigs. The pre-Darwinian classifiers were doing what seemed natural, and yet their practices were in fundamental conflict with the traditional single Chain of Being and also with Cuvier's four chains. A chain is not inherently hierarchical in the fashion of a tree. The recognition of evolution resolved this contradiction, and Darwin gave meaning to classification.

In reality, the picture is less simple than I have made it out to be. Classification of species into higher categories is somewhat arbitrary. Exactly where we draw a line between two families or genera entails the sorts of difficulties that one might encounter in dividing an actual tree branch into several natural clusters of smaller branches. Sometimes the clusters would not be so discrete that two people would assign to any one cluster exactly the same branches or even recognize the same number of clusters. The quasi-natural status that Darwin granted traditional categories of classification did not constitute compelling evidence in favor of evolution, but it did place his evolutionary scheme in a favorable light.

Stronger evidence came from embryological considerations, which had moved others before Darwin to entertain ideas of evolution, but generally with adherence to the concept of a *Scala Naturae*. Darwin cited as evidence of common ancestry the well-known fact that the early embryological stages of all vertebrates—fishes, amphibians, reptiles, birds, and mammals—resemble each other to the degree that they are difficult to tell apart. Early in the nineteenth century, the alleged appearance of a fishlike configuration in an amphibian's early development or an amphibianlike configuration in a mammal's early development was cited in a somewhat romanticized way as evidence of an animal's ascent of the Scale of Being during embryonic development. Von Baer, before Darwin's evolutionary writings, had dispelled this notion by showing that patterns of development were not so simple. A mammal, for example, does not pass through stages resembling adult amphibians or reptiles. Rather, the early development of a mammal resembles the early development of these other groups.

Darwin capitalized on the fact that the alteration of stages of development often comes late in the history of an embryo, when an animal is approaching the time when it will have to assume its specialized behavior in the outside world. In the framework of evolution, this makes sense. Presumably, evolution would have had its weakest effect on the earliest stages of development. Mammals had distant ancestors that were fishes, but why should a mammalian embryo differ greatly from a fish embryo long before birth, when each is shielded from its external environment? It was easy for Darwin to see how evolution might have left the earliest stages almost unchanged. On the other hand, why would a creating divinity force all vertebrate beings through the same series of shapes before allowing them to diverge and assume the shapes associated with their postnatal habits?

Among Darwin's specific examples was an analogy from animal husbandry. Domestic dogs, like the bulldog and greyhound, that look quite different as adults show much greater mutual resemblance at birth. (Darwin made visits to whelping bitches to make the critical observation.) The embryos of the different breeds of dog resemble each other even more closely, and, in fact, the resemblance increases the earlier the stage of embryonic development. If this pattern of divergence characterized evolution that was induced by humans through artificial breeding, was it not probable that similarly diverging patterns of development in nature reflected some naturally occurring form of evolution?

Darwin also cited as evidence structures that we would now term "vestigial." These are rudiments that seem to serve no present function but that resemble more fully developed, functioning organs in other animals. Some examples are striking. The poorly developed pelvic bones and hind limbs of certain snakes do not fit the animal's basic body plan, but can easily be accounted for as vestiges of ancestral organs not yet fully lost through evolution. The same argument applied to the pelvic bone of whales. How, Darwin reasoned, could any creator of a perfect global system have been so wasteful or imprecise in his productions?

Homology, or similarity through common ancestry, formed another part of Darwin's anatomical case for evolution. In certain animals and plants, the configurations and mechanical operations of many nonvestigial organs appeared awkward, wasteful of materials, or inefficient. Darwin noted that some of these imperfections seemed to have arisen because the structures of the organs had been restricted to certain common themes of "design." Why were the same bones present in the wing and in the leg of a bat? No omnipotent architect of life would have constrained his body plans in such a way, but evolution, forced

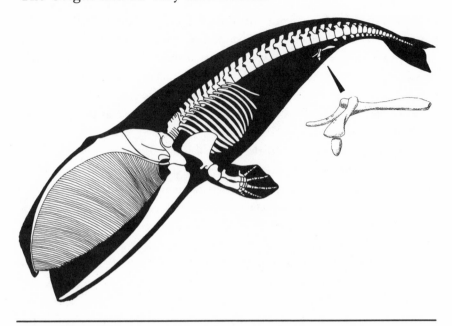

FIGURE 3.2

A baleen whale skeleton, with an enlargement of the vestigial pelvic bones.

to employ ancestral ground plans as starting points, would automatically account for the imperfections.

It might almost seem that such obvious arguments as those of the preceding paragraphs would have made earlier students of anatomy highly skeptical of creation, but there were historical reasons why this was not the case. Richard Owen, who became one of Darwin's most vicious opponents, had earlier in the century found a way of cloaking the anatomical facts in traditional raiments. He proposed the preexistence of the "archetype" for any coherent group, like the vertebrates. This was an idealized organism that possessed the fundamental features of all members of the group. Owen admitted geological history to his scheme. The member species represented the unfolding through time of the complete design for the group—the elaboration of the archetypical ideal. In effect, Owen was elevating to a higher level of classification the Platonic ideal of the species—the concept of perfection that had led to the traditional idea of the immutability of species. Owen's scheme also would explain the seemingly natural hierarchical grouping of species that was not easily reconciled with the *Scala Naturae*. Owen's scheme was heavily teleological: it implied a divine plan. The idea had general appeal because early in the nineteenth century nature was considered to manifest the Creator, and a widespread goal of science was to elicit from nature evidence of divine wisdom and purpose.

The fossil record—the preserved "broken branches" from Darwin's "great Tree of Life"—might have constituted the virtual proof of the pudding. Thus, it was naturally to the fossil record that Darwin looked for sequences of intergrading populations—the evolving lineages that I described in the first chapter—the branches of his evolutionary tree.

Unfortunately, Darwin found the record a great disappointment. Only bits and pieces of branches had come to light. One could postulate continuity between fossils that seemed to represent ancestral and descendant forms, but the record itself provided no documentation of continuity—of gradual transitions from one kind of animal or plant to another of quite different form. In the *Origin,* Darwin acknowledged the severity of this failing. His recourse was to launch a large-scale attack on the quality of the fossil record. He argued that the record was woefully incomplete and could never be expected to support his scheme of gradual and continuous change.

That Darwin was misguided in his denigration of the fossil record is the point of departure of newer arguments about evolution that I will take up in detail in chapter 5. For the present, let me point out that when we consider the limited amount of effort that in Darwin's day had been expended in the study of rocks and fossils, his arguments were necessarily speculative. They were predictions, based on what Darwin had learned about processes of geology and on what he wanted to believe about evolution; they were most definitely not conclusions derived from intensive scrutiny of the fossil record itself. In particular, what was not available for Darwin's consideration was abundant information on the geological longevities of fossil species.

What for Darwin remained of value in the fossil record were less decisive forms of evidence. One was his previously noted observation that bizarre living animals of South America, like the armadillo and arboreal sloth, had close relatives among the extinct faunas of the same continent. The same condition held for the marsupials of Australia. I have also described the importance of fossils in dramatizing for Darwin the geological fact of extinction and, more specifically, in suggesting general biotic instability and selective failure among species. Even by 1859, fossil evidence of general biotic change from the Paleozoic through the Cenozoic was too weak to have received much coverage in the first edition of the *Origin.* In subsequent editions, however, Darwin added citations to fossil data that revealed crude sequential changes, particularly for mammals of the recent geological past. Presumably, Darwin was all the more reticent on this issue in the first edition because of the opposition of Lyell, who had a powerful personal influence upon him. As I have already noted, Lyell, with all his geological expertise, continued to deny fossil evidence of succession until purely biological arguments in the first edition of the *Origin* persuaded him to abandon his antievolutionary position.

Another element in Darwin's case for evolution is frequently overlooked. This is his provision of a reasonable mechanism for change where none had been recognized. Lack of awareness of a feasible mechanism had previously been a major obstacle to belief in evolution. In fact, regardless of its effectiveness for producing large-scale change, the idea of natural selection, once unleashed by Darwin, was difficult to deny: selection had to be at least a weak modifier of species. Faced with the modest variability of form that Darwin documented for species and the fact that many more individuals were born than could survive and reproduce, biologists who read the *Origin* were hard-pressed to deny that some generation-by-generation change would occur—that some selection must take place. Darwin, at the very least, dealt a blow to the traditional belief in the "fixity" of species and, thereby, to the immutable perfection inherent in the concept of plenitude. This was a blow against antievolutionary forces in general.

Thus far, I have reviewed only the diverse lines of evidence that Darwin marshalled in favor of the reality of evolution. The mechanism, natural selection, was treated extensively in the *Origin of Species,* but the arguments supporting it were few in number. They were nonetheless powerful and cogent. Darwin seems to have recognized the importance of demolishing a troublesome barrier by convincing readers of the efficacy of his process: his revealing a reasonable but previously unrecognized modus operandi of nature would lead the skeptic toward evolution. Chapter One of the *Origin* describes variation among domestic plants and animals and outlines human success in the artificial selection. Darwin's review of the wide variety of "domestic productions" was a clever opening because what he was presenting was, in effect, a working model —a complex experiment that his predecessors had conducted for him. He could not argue that selection operated in precisely the same way in nature as in the barnyard, but he could make one very important point: sufficient variation of form could be generated within a population for selection to produce structures and habits quite different from those initially present.

What Darwin was confronting here was the venerable Platonic ideal of the species—the ideal that related to immutability. Deviations from the ideal form had been seen as unfortunate aberrations. The idea that variants must be ill-adapted and improbable seems even to have colored earlier observations of nature. Remarkably, biologists had all but denied the existence of substantial variability that was in fact fully visible. In Darwin's privately circulated "Essay of 1844," where he set forth his ideas on natural selection without going to press, he opened the discussion of natural selection with the bald apology, "Most organic beings in a state of nature vary exceedingly little." [1] In contrast, fifteen years later, in the *Origin* (p. 45), he was able to write:

> I am convinced that the most experienced naturalist would be surprised at the number of the cases of variability, even in important parts of structure which he could collect on good authority, as I have collected, during the course of years.

Darwin's shift toward the positive indicates, first, that early on he was deeply troubled by the alleged absence of variability within species and, second, that he set out to solve the problem. In the *Origin*, he documented his more optimistic claim with more examples than appear in the "Essay of 1844," and because the *Origin* was meant to be merely the abstract of a much larger work, he held other examples in reserve. Further evidence that Darwin considered artificial breeding to be valuable in providing evidence of variability is the title for his first chapter: "Variation under Domestication." He chose this title rather than "Success in Artificial Breeding" or some other heading that would have stressed selection itself.

Regardless of its title, Darwin's first chapter offers a compelling case for selection. It is difficult to emerge from a reading of this chapter without being convinced that selection, under at least some circumstances, must operate with great efficacy in nature. As Darwin recognized in the *Origin* (p. 109), one thing that nature has in far greater abundance than does human culture is time:

> Slow though the process of selection may be, if feeble man can do much by his powers of artificial selection, I can see no limit to the amount of change . . . which may be effected in the long course of time by nature's power of selection.

Extrapolating from the microcosm of barnyard selection to the macrocosm of nature herself, Darwin, by 1859, was able to offer many examples of variability within species. This was the result of two decades of searching.

It is now generally agreed that the idea of natural selection struck Darwin in September of 1838, nearly two years after the end of his voyage on the *Beagle*, when he read Malthus' well-known book on population growth (the book at that time was forty years old). Darwin had believed in evolution for more than a year and was searching for a mechanism. He found it in the Malthusian notion that numerous offspring die for every one that survives. Darwin recorded the moment of enlightenment in his autobiography:

... fifteen months after I had begun my systematic enquiry, I happened to read for amusement Malthus on Population, and being well prepared to appreciate the struggle for existence which everywhere goes on from long-continued observation of the habits of animals and plants, it at once struck me that under these circumstances favourable variations would tend to be preserved, and unfavorable ones to be destroyed. The result of this would be the formation of new species. Here, then, I had at last got a theory by which to work.[2]

Darwin's notebooks for this period have only recently come to light. They were published in 1967. Sandra Herbert has called attention to a critical passage that identifies September 28 as the actual day when Darwin recorded his seminal thoughts on natural selection:

Population is increase [d] at geometrical ratio in *far shorter* time than 25 years—yet until the one sentence of Malthus no one clearly perceived the great check amongst men. . . . Take Europe on an average every species must have same number killed year with year by hawks, by cold &c.—even one species of hawk decreasing in number must affect instantaneously all the rest—The final cause of all this wedging, must be to sort out the proper structure, & adapt it to changes. . . . One may say there is a force like a hundred thousand wedges trying force every kind of adapted structure into the gaps on the economy of nature, or rather forming gaps by thrusting out the weaker ones.[3]

It is still debated whether Darwin's reading of Malthus effected a conversion as instantaneous as St. Paul's on the road to Damascus, or whether it produced the last step in an accumulation of reasoning. In any event, as Sandra Herbert has observed, a reading of Malthus inspired Darwin to transfer the idea of a struggle of existence from one level of biological organization to another. Instead of seeing the struggle as concerning only the fate of species, Darwin saw it as concerning the fate of individuals. By the differential success of individuals, species themselves could be transformed.

A phenomenon that almost certainly emerged secondarily in Darwin's mind was the concept of sexual selection. This process would seem to be obvious once one has recognized the phenomenon of natural selection, but represents an apparent contradiction in that it produces structures that may not aid an animal in survival. Sexual selection is a form of natural selection that involves only differential fecundity. Why is a male peacock so decoratively garbed? Why do males of some hoofed species of mammals grow antlers or horns that they almost never use in warding off predators? These structures are for display or

FIGURE 3.3

Ornamentation of the male hummingbird *Spathura underwoodi* (right), shown with the female. Darwin noted that males of nearly all hummingbird species exhibit visual products of sexual selection.

rutting warfare. Males who have been particularly well endowed with them have preferentially received the favors of females or have defeated other males in contests of battle or display and have taken females for themselves. In time, incipient colors or shapes that have been favored in the pursuit of females prevail, and eventually conspicuous or even bizarre "secondary sexual adaptations" develop. Of course, if these structures reach the point of inhibiting feeding, defense against predators, or other necessary functions, they may cease to be further accentuated by sexual selection. Females as well as males may exhibit sexually selected structures. As seemingly wasteful features from the standpoint of adaptation, these structures are hardly consistent with an economical divine plan of creation. Natural selection readily accounts for their paradoxical presence, and as Darwin recognized, they represent some of the most convincing illustrations of the process.

FIGURE 3.4
Male argus pheasant, displaying its plumage during the breeding season. Darwin observed that the "eye spots" on the wings are shaded in such a way as to appear three dimensional. The plumage is clearly the result of sexual selection.

FIGURE 3.5
Face of the male mandrill, which bears bright red, white, and blue markings. Darwin judged this to be the most extraordinary coloration found among all the mammals.

I have said a great deal about what is in the *Origin of Species*. It may seem surprising that two things are not to be found there. Nowhere in the body of the *Origin* is the transformation of species labeled "evolution." In Darwin's day, "evolution" referred to the unfolding of the individual during its development. "Evolved" appears only as the last word of the *Origin*. Little did Darwin know that here he fixed the word's subsequent meaning. Similarly, the familiar phrase "survival of the fittest" is not present in Darwin's great book. This phrase that caught the public's fancy is Herbert Spencer's. In correspondence, Darwin professed a liking for the phrase, but found its chief shortcoming to be that it could not serve as a verb for describing the evolutionary process. Finally, Darwin avoided any discussion of a role for evolution in human origins. The basic idea of transformation by natural selection was a large enough burden to heap suddenly on the shoulders of an unsuspecting world. Treatment of human evolution would have to be deferred. In 1871, Darwin published *The Descent*

The *Origin* and Its Very Slow Process

of Man and Selection in Relation to Sex.[4] The book's title identified two topics unlikely to be warmly welcomed in the chilly Victorian age.

For two decades, from 1838 to 1858, Darwin delayed the publishing of his ideas on evolution by natural selection. As early as 1842, he set down the rudiments of his argument (the "Sketch of 1842"), and two years later he spelled it out more fully in a second manuscript (the "Essay of 1844"). His wife was instructed to publish this later document in the event of his death. Thus, he secured his priority and ensured that his conclusions would not be lost to science.

Darwin's delay seems linked by common causes to his famous retreat from public life. What deserves more immediate consideration is the particular aspect of Darwin's work now under challenge. Why was he so thoroughly wedded to a gradualistic view of evolution? Why did he reject out of hand the alternative (which I will argue the fossil record now favors) that evolution moves rapidly in small populations, when certain species originate, while for most populous, well-established species, transformation is remarkably slow? In his recapitulation and conclusions, which formed the final chapter of the *Origin,* Darwin summed up the temporal dimension of his scheme (p. 471):

> As natural selection acts solely by accumulating slight, successive, favourable variations, it can produce no great or sudden modification; it can act only by very short and slow steps. Hence the canon of "Natura non facit saltum", which every fresh addition to our knowledge tends to make more strictly correct, is in this theory simply intelligible.

The Latin phrase meaning "Nature does not make jumps" was a venerable dictum associated with the *Scala Naturae* which, as an embodiment of perfect divine planning, could harbor no gaps. It is a measure of Darwin's desire to underscore slow, continuous modification that here he violated his own philosophy of empiricism and reached back into history for what was essentially religious dogma. Certainly there was no factual evidence that gradual, persistent evolutionary trends formed the tree of life. Documentation could have come only from fossils, yet Darwin believed that the record sealed within the rocks was too fragmentary to settle the question.

It was in part the new uniformitarian geology that fostered Darwin's inclination toward gradualism. Citing Lyell in the *Origin* (p. 95), he compared the evolution of life to the slow, piecemeal reshaping of the Earth's surface—the conception of Earth history that had displaced catastrophism:

Natural selection can act only by the preservation and accumulation of infinitesimally small inherited modifications, each profitable to the preserved being; and as modern geology has almost banished such views as the excavation of a great valley by a single diluvial wave, so will natural selection, if it be a true principle, banish the belief of a continued creation of new organic beings, or of any great and sudden modification in their structure.

It is true that the new geology made feasible the notion of gradualistic evolution, but it in no way excluded the possibility of a more pulsating tempo. There is abundant evidence that other factors influenced Darwin in his total denial of rapid evolutionary steps.

One probable influence was Darwin's evident fear that his ideas would not be widely accepted—a fear that is well expressed in his correspondence. He chose to argue that natural selection was an excruciatingly subtle process that could be weak yet effective because of the availability of vast stretches of time. This would clearly be more palatable to others than the alternative: that the process wrought changes rapidly. How else could his colleagues accept the requirement that the process had gone on before their very eyes without their having noticed? Darwin, in fact, had an exaggerated view of geological time. He estimated that the erosional denudation of the British Weald Formation had occupied 300 million years. The exact geological interval to which Darwin was referring is unclear, but his estimate of the actual time represented was excessive by a factor of at least three and possibly four or five.

Presumably another influence upon Darwin was the natural tendency to polarize: we tend to push a new idea of our own toward one extreme, in the process of contrasting it with competing notions. For Darwin, the alternative mechanism to be displaced was Divine Creation. If he had argued instead for something akin to the modern punctuational model, he would have been offering something that was no more cogent in a scientific sense. Nothing was then known of genetics or of the many other sources of information that today make the punctuational view biologically feasible. Had Darwin claimed that natural selection operated most effectively in small populations so that we should not expect to observe its major effects directly in the modern world or indirectly in the fossil record, his arguments would have lost credibility. Selection would have become a phantom process, doing most of its work exactly where we could probably never observe it. Although natural selection had not been recognized in the sizable populations upon which Darwin focused, at least these were large, tangible entities—including all of the species of the world—and they could be studied in the future for evidence of past or current transformation.

The *Origin* and Its Very Slow Process

It seems clear that strictly biological concerns also contributed greatly to Darwin's gradualism. One was the continued prevalence of the belief that many organs were complex yet also perfectly adapted. In his "Sketch of 1842," under the heading "Difficulties on Theory of Selection," Darwin jotted, "It may be objected such perfect organs as eye and ear, could never be formed. . . . But think of gradation. . . ."[5] This problem was, in fact, the only "difficulty on the theory" that he discussed. Thus, he apparently saw as a major obstacle the well-entrenched belief in perfect adaptation that Cuvier had successfully preached. This concept was rooted in the Christian notion of plenitude that had dominated eighteenth-century natural philosophy. Darwin had to approach the question of perfection gingerly. He had to deny absolute perfection and yet give selection the opportunity to mold intricate structures that were remarkable, if not absolutely perfect. Natural selection was a perfecting mechanism, but organisms could not at any time be perfect or they would not be evolving. Even in a changing environment, some lag time was to be expected between external demands and evolutionary response. "The wonder indeed is, on the theory of natural selection, that more cases of the want of absolute perfection have not been observed," Darwin declared in the *Origin* (p. 472). But he was bucking the tide of history and he knew it. The idea of a perfect structure appearing full blown was entirely unacceptable, and he removed himself as far as possible from this alternative by accepting a perfecting process that was almost imperceptibly slow. He felt compelled to give evolution as much time as possible to work its wonders, and Lyell's gradualistic picture of Earth history had provided him with the time he needed.

I have previously noted the great irony that Darwin named his book *On the Origin of Species,* but had little to say about how species actually multiply. He was primarily concerned with the gradual transformation of a species to the point where it would be called something else. One important matter on which he was equivocal was the role of geographic isolation in the branching off of new species. The process of geographic speciation, as championed by Ernst Mayr in this century, is now widely believed to be a prevalent mechanism by which species multiply. The theory here is that the new species evolves from one or more populations that have become separated from the ancestral species by some sort of geographical barrier. Biologists, including Mayr, who have concluded that rapid evolution sometimes occurs during the branching process have envisioned such rapid transformation as occurring only when the seminal population is small (and even then, often only weak transformation is effected). It is currently debated how frequently a small population may diverge rapidly to form a new species without geographic isolation. What is important, however, is population size. For rapid change to occur—for Darwin's slow tempo to be violated—a small population is usually required. As I will more fully outline in chapter 5, it is only in a small population that unusual new features can easily be emplaced. One reason for this is that with few breeding individuals, complete mixing of the gene pool is possible. To take an extreme case, if there are only four individuals in a population, and if one of the four possesses a unique but highly valuable trait, there is a good chance that this trait will be able to spread throughout the population during a few generations of subsequent evolution, provided that the population does not grow appreciably. It is much less likely that such a trait will spread rapidly throughout an enormous, widely dispersed and fragmented population—throughout a typical large and well-established species. Because of their size, small populations also tend to occupy discrete habitats. If one happens to occupy an unusual habitat that imposes natural selection strongly in a particular direction, that population may change quickly. Far less likely is the possibility that a large, widely dispersed population occupying a variety of habitats will be subjected to strong, consistent selection—that it will move very rapidly in any particular direction.

Having a poorly developed notion of speciation, which in the punctuational view of evolution is the critical process, Darwin failed to focus upon the role of geographical isolation or, more generally, upon the importance of evolution in very small populations. His posture here is puzzling for two reasons. One relates to the fact that throughout his life as an evolutionist, he had no access to modern genetics, a field which was not born until the start of the twentieth century. Darwin, who died two decades earlier, remained saddled with the contemporary concept of blending inheritance. This meant that in his day discrete units of genetic inheritance were unrecognized. It was not appreciated that an inherited aspect of form or behavior could go unexpressed for generations and then fully reappear, having lain dormant because of being governed by a recessive gene, so that it only emerged if such a gene happened to be inherited from each parent. The idea of blending inheritance followed from the seemingly reasonable, yet often incorrect, assumption that any trait of an individual was essentially the average of the comparable traits of that individual's parents. Certainly, it could be seen that offspring did not grow to heights exactly intermediate between the heights of their parents, but this was not a fatal observation because the degree to which visible traits were inherited was uncertain. An unusually tall child might merely be the result of better nutrition. Given a false belief in blending inheritance, it becomes difficult to imagine how even valuable new adaptations can come to prevail; with every breeding they will be averaged with others. How could selection possibly overcome such dilution instead of becoming blended into oblivion? Darwin was vigorously attacked on this point late in his life and was deeply disillusioned by his inability to extricate his theory from the problem. He could have escaped by focusing on evolution within very small populations isolated by geographic or reproductive barriers. Here, even with blending inheritance, inbreeding could have fixed unusual features. Unfortunately, Darwin apparently overlooked this crucial point.

The second reason that we might expect Darwin to have appreciated the role of small, localized populations is that much of his evolutionary insight came from the study of island biotas. How many individuals could have founded each of the species of bird unique to the Galápagos Islands? In some cases, probably only a single breeding pair. In others, not more than a handful of birds. Darwin had witnessed results of the remarkable divergence of small isolates, but he constructed a theory that he applied primarily to the gradual transformation of entire species.

Thus, it is paradoxical that while brilliant geographic observations molded Darwin's belief in the reality of evolution, in the eye of the modern punctuationalist, his greatest error was also in this area. Once Darwin conceived of an evolutionary mechanism, he failed to give it the particular geographic focus that would have rendered it most effective.

Another exceedingly important determinant of Darwin's gradualism was his belief that variability among organisms—the raw material of selection—was a sparse commodity in nature. I have already documented this point, including the fact that in 1844 he believed variability to be virtually absent—"probably quite wanting (as far as our senses serve) in the majority of cases," he wrote.[6] He spent much time during the next fifteen years gathering examples of variation within species, but throughout this interval he retained the notion that cultivated plants and domesticated animals exhibit greater flexibility of form than is found in nature. This we see in the very first sentence of Chapter One of the *Origin:*

> When we look to individuals of the same variety or subvariety of our cultivated plants and animals, one of the first points which strikes us, is that they generally differ much more from each other than do the individuals of any one species in a state of nature.

Variability became an obsession with Darwin. In 1868, he published his enormous two-volume monograph, *The Variation of Animals and Plants under Domestication*. It is quite evident in the *Origin* that he saw weak variability in nature as potentially being a major barrier to the efficacy of selection (p. 108):

> That natural selection will always act with extreme slowness, I fully admit. . . . Nothing can be effected, unless favorable variations occur, and variation itself is apparently always a very slow process.

In part, Darwin was laboring under a burden with which he was saddled by his predecessors, his work having been undertaken at a time when, through retention of the Platonic ideal, biologists saw species as having precise, perfectly adapted shapes. The study of species was "typological": one good specimen served to define a species. In time, the *Origin of Species* righted this absurd, nonempirical tenet of the old biology, but this happened after the fact. Variability had not been studied formally because it had been denied a priori. Darwin made headway in collecting examples after his "Essay of 1844," but he remained apologetic. Furthermore, he had to convince others, and this was even more difficult. In the *Origin* (p. 45), he acknowledged, "It should be remembered that systematists are far from pleased at finding variability in important characters."

A related concept of the traditional biology also confined Darwin to a gradualistic path. This was the idea of plenitude. As we have seen, Lyell and others had perceived a natural "struggle for existence," with species appearing and disappearing, but they also believed that biotic saturation was maintained. Species that disappeared were replaced by new ones or by old ones that expanded to assume vacated roles. Darwin proposed that species were modified, but he also clung to a belief in the balance of nature or, as he called it, the "economy of nature." In doing so, he was confronted with a paradox. How could species vary significantly in a world into which they were packed like sardines? Here we can recall from his notebook of 1838 the simile of his very first description of natural selection—"a force like a hundred thousand wedges trying to force every kind of adapted structure into the gaps in the economy of nature, or rather forming gaps by thrusting out weaker ones. . . ."[7] We see that at the very moment of germination of his great idea, the "Catch-22" surfaced: selection operates by sifting through variations among individuals, but life is so tightly packed, according to the economy of nature, that exceedingly little variability is tolerated. Darwin's first clause postulated gaps for the "wedges" to fill, but then he caught himself. No, the wedges had to make their own gaps. This could not be easy under conditions of plenitude. Darwin, like modern biologists, saw the variation upon which selection operated as being accidental, or random, in origin. Given the close confinement and near perfection that Darwin envisioned for each species, would not almost all accidental variations be deleterious?

In other words, Darwin's rigid belief in plenitude gave his process precious little room to operate. Unfortunately, variation, the basis for all change, had to be held to a minimum. He could squirm out of such tight quarters only by suggesting that minor perturbations might be propagated throughout the system like gentle waves. He suggested in the *Origin* (p. 108) a subtle domino effect: "But the action of natural selection will probably still oftener depend on some of the inhabitants becoming slowly modified; the mutual relations of many of the other inhabitants being thus disturbed." He followed this passage with a reiteration of his unequivocal claim that evolution always proceeds with "extreme slowness."

Clearly Darwin's adherence to the doctrine of plenitude was a prime reason why he could visualize only gradual change. Possibly we also have here an explanation for his difficulty in coming to grips with the rapid origin of species from small populations. Under conditions of global plenitude, how could he easily have imagined that a small population might expand to take firm hold of a place in nature?

Now it appears that we have a way out. In the chapters that follow, I will offer evidence that the concept of plenitude is unfounded. The world is packed loosely enough with species that much variation is tolerated and occasionally small populations do diverge rapidly into distinctive new species that multiply and prosper. There is, in fact, an "experimental" quality about the way in which evolution proceeds on a global scale. Selection does operate, but in a much less constrained manner than Darwin was forced to believe.

4

Darwinism Challenged, Darwinism Affirmed

DARWIN'S BOOK shook the world in 1859 and his arguments are widely accepted in our time, but the fortunes of his ideas have nonetheless followed an irregular course during the twelve decades in between. Evolution per se has fared well throughout, but the concept of natural selection has had its ups and downs. Until shortly before 1940, biologists who accepted Darwin's mechanism as the dominant source of evolutionary change were in the minority. Certainly, among Darwin's contemporaries there were few ardent selectionists.

Darwin employed certain strategies in attempting to convert the world to his beliefs. Perhaps unconsciously, he practiced a simple procedure to disarm his partial opponents: he claimed them as partial supporters. His correspondence reveals his habit of flattering and lavishing appreciation on those who expressed some agreement. His answers to their letters about the *Origin* conveyed both

Darwinism Challenged, Darwinism Affirmed

his strong desire for their loyalty and his great sensitivity to criticism. This had the effect of crytallizing support: for the respondents subsequently to weaken their enthusiasm or reassert their negative attitudes became an aggressive act. Hugh Falconer's work on fossil elephants, which I will describe in the following chapter, gave Darwin some cause for concern. Falconer's findings could have led Darwin toward the punctuational idea that many species evolve very rapidly when they first form, while nearly all well-established species evolve quite slowly. Darwin's approach, however, was to appease rather than do battle or substantially restructure his original views. He addressed Falconer's letter and potentially problematical manuscript in saccharine tones:

> Your note and every word in your paper are expressed with the same kind feeling which I have experienced ever since I have had the happiness of knowing you. . . . There's not a single word in your paper to which I could possibly object: I should be mad to do so; its only fault is perhaps its too great kindness.[1]

Only after these softening remarks did Darwin confront the issue: "Your case seems the most striking one which I have met with of the persistence of specific characters." (Falconer claimed that species of fossil elephants persisted for long intervals with little evolution.) Three years earlier, Darwin had sent Falconer a complimentary copy of the *Origin* with another deferential letter:

> Lord, how savage you will be, if you read it, and how you will long to crucify me alive! I fear it will produce no other effect on you; but if it should stagger you in ever so slight a degree, in this case, I am fully convinced that you will become, year after year, less fixed in your belief in the immutability of species. With this audacious and presumptious conviction,
>
> I remain, my dear Falconer,
Yours most truly,
Charles Darwin[2]

It is hardly surprising that Falconer, unlike some other students of fossils, did not oppose Darwin altogether, but instead proposed that natural selection might indeed have produced the species of fossil elephants in question, but during a very brief interval of time.

Darwin more consciously followed a second strategy to promote his ideas. This was to win over three prominent men whose influence would impart momentum to the cause: Charles Lyell, Joseph Dalton Hooker, and Thomas Henry Huxley. They were, respectively, Britain's most formidable geologist, botanist, and zoologist-paleozoologist. The way in which the three responded and interacted with Darwin reveals much about Darwin himself and about the

55

FIGURE 4.1
Charles Darwin in 1854, five years before publication of the *Origin*.

way in which many of his ideas became assimilated into the general corpus of biological thinking.

Lyell, of course, had been well aware of Darwin's ideas before 1859 and, though not well disposed toward them, had sufficiently appreciated their great significance that he repeatedly urged Darwin to publish without further delay. Lyell was sixty-two years old when the *Origin* appeared, and it is a tribute to his flexibility, as well as to the power of the *Origin,* that, as a staunch believer in the immutability of species, Lyell eventually tilted toward evolution. Lyell, like nearly all others, was greatly impressed with Darwin's reasoning and documentation, but while he thought the argument sound, he remained cautious. He was simply too set in his ways for rapid conversion, and in his book *Geological Evidence of the Antiquity of Man,*[3] while he entertained the possibility of an evolutionary origin for the human species, he took no positive stand. Darwin thought that he had earlier received indications of greater support from Lyell and was disappointed that Lyell, in the great geologist's own private phrase, would not go "the whole orang." Lyell did declare himself for evolution in the tenth edition of his *Principles of Geology* (1867), but never satisfied Darwin with a firm commitment to natural selection. On the other hand, Lyell bolstered Darwin's case indirectly. Having for years condemned the fossil record as being too incomplete even to illustrate large-scale biotic change, Lyell now claimed that the record was clearly inadequate to the task of documenting the validity of the theory of evolution on a finer scale. Here Darwin found support for his claim that uncooperative fossil data could not erode his gradualistic position.

Hooker was by no means shocked by the *Origin* because he had fifteen years earlier read and commented upon Darwin's unpublished "Essay of 1844." He published a review of the *Origin* in December of 1859. His stance was essentially neutral: evolution by natural selection stood as a solid hypothesis in competition with Divine Creation. A year later, Hooker described himself as an "unwilling convert," and he seems by this time to have been more fully converted than Lyell and Huxley were ever to be.

Huxley was a man who by nature opposed convention, and it was partly for this reason that he became Darwin's most vocal advocate or, as he was known, "Darwin's Bulldog." He saw Darwin's work as masterful, provocative, and deserving of fair consideration. Huxley became a wholehearted evolutionist, but remained skeptical of natural selection. He was troubled by a pair of related matters, both of which I will examine more fully in the following two chapters. One was Darwin's utter rejection of the possibility of rapid evolutionary steps. The other was Darwin's acceptance of the great geological antiquity of many living groups of animals. Huxley noted that some groups of modern mammals, for example, extend far back into the geological record. He proposed that for Darwin's gradualistic scheme to be valid, the origins of many other advanced groups must lie much farther back in the record than seemed apparent. Alternatively, he reasoned, evolutionists must admit the importance of rapid steps. Even to accept the validity of evolution in general, Huxley demanded firm fossil evidence. On this point, at least, he eventually became convinced.

What is remarkable about Darwin's primary triumvirate of defenders was that they, especially Lyell and Huxley, became his defenders without fully accepting his conclusions. This was a new kind of development for biology. So cogent were Darwin's arguments and so abundant were his facts that he was lionized for scientific technique, not for ultimate truth. Huxley's reaction to the concept of natural selection was, "How extremely stupid not to have thought of that," and Hooker confessed to Darwin that he himself would rather have written the *Origin* than any other book. Here they were praising the insight and methodology of Darwin's science rather than his particular results. Their reaction indicates how sharply Darwin's reliance on fact and logic stood in contrast to the speculative approaches to biological theory that had preceded him.

Soon after seeing the *Origin* published, Darwin withdrew from the debate over its contents. In July of 1860, he wrote to James Dwight Dana of Yale, "On principle, I have resolved to avoid answering anything, as it consumes much time, often temper, and I have had my say in the *Origin.*"[4] He continued to publish on related matters, but with regard to evolution per se, he relied on others, most notably Huxley, to do his bidding.

Darwin's twenty-one-year postponement in going to print with his ideas, like Hamlet's famous delay, has been the object of much analysis. Darwin's reluctance obviously had many of the same causes as his personal retreat after publication. He was, in fact, confronted by real problems. It appears that during his years of delay, he was justifiably worried by two practical matters. These can be surmised from what he undertook to accomplish during the delay. The first difficulty, which many historians have recognized, was his need to establish a reputation in empirical science: he who would write authoritatively about the transformation of species must become expert in the biology of species. The obvious avenue here was to specialize on a particular taxonomic group. Partly because of experience with barnacles on the voyage of the *Beagle,* Darwin undertook to study and classify these small crustaceans, most of which lie upside down in shells and kick food into their mouths with their feet. Barnacles not only encrust the hulls of ships, but also blanket rocky shores and attach to whales and other animals. From 1846 until 1854, Darwin worked intensively on living and fossil barnacles. His monographs are still considered to be of high quality, and the research gave him insight into the philosophy of classification that he introduced in the *Origin.* "Nonetheless," he wrote in his autobiography, "I doubt whether the work was worth the consumption of so much time."[5] While conducting the tedious barnacle studies, however, Darwin was fully aware of the opprobrium that had been heaped upon speculative evolutionists who preceded him. One of these was Robert Chambers who, lacking scientific credentials and attempting to maintain anonymity, had published an insolent and superficial evolutionary book, *Vestiges of Creation.*[6] This inflammatory volume had appeared in the very year when Darwin finished the "Essay of 1844," which he squirreled away to be published only in the event of his death.

The other genuine problem confronting Darwin was the one I have already noted, but one that historians seem generally not to have recognized. Darwin was compelled to demonstrate the existence of variation within species—variation upon which selection could act. During the 1840s and 1850s he conversed and corresponded extensively with animal and plant breeders and undertook his own breeding experiments, especially with pigeons. Darwin's preoccupation with variability in form and behavior is evident in his major publications. When he interrupted the writing of his magnum opus on evolution to hurry to press with the *Origin,* he had completed preliminary drafts of eleven chapters of the incipient larger work. These were condensed to form the text of the *Origin,* but it is notable that the only two of the large preliminary chapters that he later chose to develop fully as a separate book were the chapters on variation. These became his *Variations of Animals and Plants Under Domestication,*[7] published in two volumes in 1868. Darwin's preconception had been that species varied quite little in nature. Here he simply adopted the standard typological myth dictated by the almost universal belief in species' immutability. A telling passage appears in a letter of June 13, 1849, addressed to Hooker; Darwin's subject is his barnacle work:

> I have been struck . . . with the variability of every part in some slight degree of every species. . . . I had thought that the same parts of the same species more resemble (than they do anyhow in Cirrepedia) objects cast in the same mold. Systematic work would be easy were it not for this confounded variation, which, however, is pleasant to me as a speculatist. . . .[8]

What he meant by "speculatist," of course, was theorist in the area of natural selection. Only by poring over small barnacles, conducting his own breeding tests, and accumulating a wealth of information from other breeders was Darwin able to overturn the reigning typological dogma and proclaim the variability that his theory demanded. All this took time.

Darwin's providence in assembling a powerful case was well founded, but so was his continued anticipation of criticism. In the period before publication, two of his three primary targets, Hooker and Lyell, had been subjected to his evolutionary propositions for substantial periods of time without full acceptance. Inevitably, his more general reception would also be mixed.

Not surprisingly, a trait common to nearly all who fervently resisted the arguments of the *Origin* was a firm religious conviction. Because science and religion had been tightly intertwined, the reaction of theologians was not without impact. The most aggressively antagonistic was Bishop Wilberforce of Oxford, famous for challenging Huxley in formal debate as to whether he was descended from a monkey on his grandfather's or grandmother's side. The rapier-tongued Huxley muttered aside, "The Lord hath delivered him into mine hands," and proceeded to decapitate the bishop verbally.

Christianity had reached an accommodation with Lyell, but Lyell had required of religion only a withdrawal from the literal interpretation of the biblical account of the Earth's physical history. Life was far more sacrosanct than rocks and oceans—particularly human life. Although Darwin had carefully avoided broaching the subject of human origins, the implications of his story were obvious. As a result, Darwin's book quickly brought about a feeling of spiritual isolation among his contemporaries. Here the *Origin* was a kind of last straw. Long before, Copernicus had removed the Earth from the center of the universe and Halley had set it afloat in a giant galaxy. Now Darwin was demoting humans from status as God's favored creatures on one small planet to the level of brutes, as nonhuman mammals were customarily called.

The reception of the *Origin* is difficult to summarize. Perhaps the simplest summary point to be made is that nearly all natural scientists were highly impressed with the book, but while the reality of evolution quickly gained widespread acceptance, the mechanism of natural selection did not. In America, Germany, and Russia, the response was generally as positive as in Britain. In France, partly for reasons of chauvinism and partly for reasons of history (Cuvier's lingering influence), the reception was lukewarm. By 1900, most French scientists were *transformistes,* but gave Darwin little credit for their conversion.

In 1880, Huxley in his essay, "The Coming of Age of the Origin of Species," was able to write of Darwin's book that "the foremost men of science in every country are either avowed champions of its leading doctrines, or at any rate abstain from opposing them."[9] His more specific comments reveal that he was referring here to the reality of evolution, not to natural selection.

Darwinism Challenged, Darwinism Affirmed

During the 1880s, the general acceptance of the concept of evolution continued, but there was a growing disillusionment with the way it was being studied and related to particular biological problems. For one thing, beginning in the 1860s, many biologists settled into the new evolutionary era with the unimaginative conviction that the most important task at hand was to delineate the shape of the tree of life. They attempted to establish the phylogeny (branching pattern) for one phylum or class after another. Conjecture ran rampant, partly because fossil data were frequently unavailable.

Other serious problems also caused a flagging of interest in evolution. Even Darwin himself was embattled, and he vacillated with indecision before his death in 1882. He faced two particular quandaries. The physicists inflicted one of these. Lord Kelvin and his followers estimated the Earth to be shockingly youthful. They founded their calculations on the assumption that the planet must be cooling down from a high initial temperature of formation. The only reason that it could now be as warm as it is—temperature increases with depth in a mine shaft—is that it is so young as to retain some of its primordial heat. More precisely, from its likely temperature at formation and estimated cooling rate, the Earth was calculated to have an age of something like twenty million years (some physicists allowed as much as forty million). The prestige of mathematical science lay behind these calculations. Kelvin revealed the arrogance of an eminent mathematical theorist trifling with an inferior science in the very title of his address of 1865: "The Doctrine of Uniformity in Geology Briefly Refuted."

Could James Hutton's venerable observation that the rocks showed "no vestige of a beginning, no prospect of an end"[10] be totally wrong? Could the accumulated evidence of many decades after Hutton be summarily discarded? Kelvin and his followers pressed their case for many years and, while not every biologist or geologist was swayed, the effect was to cast doubt upon gradualistic evolution. There was a solution to the geologists' and evolutionists' problem, but it came to light only after the turn of the century, when radioactivity was discovered in terrestrial rocks. Radioactive isotopes, by definition, decay to other isotopes, and as they do they emit heat. The Earth itself is an enormous isotopic furnace. As decay proceeds, the furnace winds down, but at the present time an enormous amount of heat is still being liberated. Lord Kelvin's calculations inevitably failed to include radioactive heating and were therefore meaningless. He ascribed to the Earth less than 1 percent of its actual age!

Darwin's second thorny problem was with heredity. In the *Origin* (p. 13) he lamented: "The laws governing inheritance are quite unknown." Throughout his life, Darwin was saddled with the contemporary notion of blending inheritance. As I have previously explained, this was the idea that two parents' traits were averaged to form the genetic endowment of their offspring. Obvious departures were assumed to reflect environmental influences. Even today the "nature versus nurture" debate continues with regard to certain features, and in the last century virtually nothing was known about the degree to which conditions of life, rather than heredity, guided the details of growth and development. It was argued, in particular by the engineer Fleeming Jenkin in 1867, that blending inheritance posed fatal difficulties for natural selection. Any new structure of some value would pass into oblivion through generation-after-generation dilution with other structures, that, because they were present earlier, prevailed numerically within the population.

Inheritance became an obstruction to Darwinism because of a misconception that, like the one pertaining to radioactivity, was not exposed until the turn of the century. The belated genetic revelation was that what really obtains is particulate inheritance. There is some averaging of the *expression* of certain genetic components inherited from parents, but the components themselves (genes) remain discrete. Genes may be expressed weakly or not at all in certain individuals, depending upon what other genes occur with them. Still, barring mutation, they persist, so that then can be fully expressed in bodily form or function in future generations. Thus, potentially valuable genetic changes are not inevitably lost through blending.

Unfortunately, lingering doubts engendered by the apparent problem of blending inheritance accompanied Darwin to his grave. The great irony was that, years before, Gregor Mendel, an Austrian monk, had in the garden of his monastery conducted brilliant experiments lucidly demonstrating particulate inheritance in the kinds of plants that bear edible peas. When Mendel read his results before the scientists of the Brunn Society in 1865, there was no response. His work was of a statistical nature and was simply not understood. Mendel was ahead of his time. His contribution languished in obscurity until the turn of the century, when its rediscovery signaled the birth of modern genetics. Mendel was well aware of the evolutionary implications of his work. As an unworldly monk, he simply failed to assert himself after a first attempt at communication. He had, in fact, sent a copy of his published paper to Darwin, but to no avail. To this day, the pages of that copy remain uncut.

Embattled as he was late in life, Darwin hedged a bit about the efficacy of natural selection. In his *Descent of Man* he wrote: "I now admit . . . that in the earlier editions of my 'Origin of Species' I perhaps attributed too much to the action of natural selection or the survival of the fittest."[11] Here and in passages

Darwinism Challenged, Darwinism Affirmed

added to the later editions of the *Origin*, Darwin suggested that he had originally gone too far in assuming that every biological feature is adaptive. He suggested that some variation that appeared was retained by mysterious "exciting causes," which were essentially accidental.

During the last two decades of the nineteenth century, the problems that beset evolutionary studies strengthened a movement toward experimental biology. It seemed that here, in imitation of the physicists and chemists, the biologists might move forward and gain dignity.

One issue that eventually fed into the new drive toward experimentalism was whether variation within species was of a continuous or discontinuous nature. The question was really one of degree, in that seemingly continuous variation is actually discontinuous. There would seem, for example, to be a continuous spectrum of heights in humans, but actually no individual is precisely as tall as any other. What the proponents of discontinuity argued, however, was the importance to natural selection of much more discrete polymorphism (literally "multiple forms"). Blending inheritance was at the root of the problem. Those who believed that selection operated on subtle variation were strict followers of Darwin (recall his emphasis on evolution by infinitesimal steps). Those who believed that more pronounced variation was most important saw it as a way out of the blending problem; some, like William Bateson, actually believed that subtle, gradational forms of variation were not even inherited.

Thirty-five years after Mendel's publication and sixteen years after his death, Mendel's work entered into the debate. In 1900, by a remarkable coincidence, Hugo De Vries, Erich von Tschermak, and Carl Correns all happened upon Mendel's long lost paper. All three were studying inheritance and recognized immediately the significance of Mendel's findings, including the most fundamental point, that traits are inherited by way of discrete units that themselves are not blended even though their effects are sometimes diluted by other genetic traits.

Apparently Mendel was judicious in choosing the pea plant for his work, because this species happens to possess a number of characters that exist in discrete, alternative forms. One of the characters is the color of the flower. Mendel found that by crossing a pure-bred strain having red flowers with another having white flowers, he obtained a second generation of plants all having red flowers (rather than the pink flowers that blending inheritance would have predicted). When these were intercrossed to produce a third generation, however, about one-quarter of the plants grew white flowers. Mendel

63

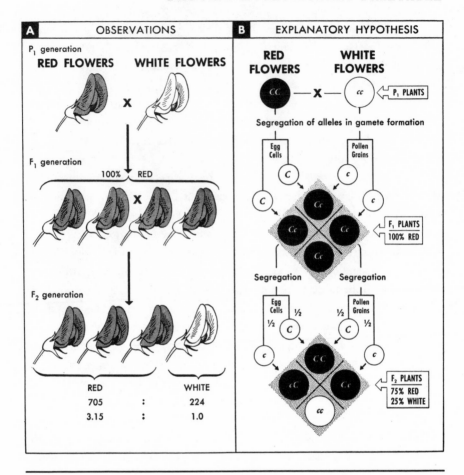

FIGURE 4.2

Mendel's results in the experimental breeding of peas. As shown in *A*, first Mendel crossed a white-flowered strain with a pure-breeding strain that had red flowers. The result was a generation (F_1) of red-flowered offspring, each of which had inherited one red and one white genetic factor, with the red one being dominant. The random allocation of the F_1 genetic factors to the next (F_2) generation resulted in about one-quarter of the F_2 individuals bearing white flowers because they had two genetic factors for these (*cc*).

recognized that the white flowers' skipping a generation was significant and so was their particular frequency when they reappeared. The same pattern turned up in crosses involving tall and dwarf plants, round and wrinkled seeds, green and yellow pods, and several other alternative characteristics. Mendel reasoned that in the initial red-white cross, although no white-flowered progeny appeared, some genetic character had been perpetuated cryptically, because white

Darwinism Challenged, Darwinism Affirmed

flowers emerged again in some members of the third generation. Something was being passed along from each original, pure-bred parent plant, but the determinant of red color, whatever it was, prevailed in the first hybrid off-spring. (We would now say that there were two alleles of the gene determining flower color, and the "red" allele was dominant, while the "white" allele was recessive.) Mendel cleverly deduced how the third generation developed. Because each member of the second generation had an allele for red from one parent and an allele for white from the other (we would now say, was heterozygotic), when these were crossed, roughly one-half of the resulting third-generation plants were similarly heterozygotic (but had red flowers because of dominance); roughly one-quarter were homozygotic red (had two "red" alleles), and roughly one-quarter were homozygotic with two "white" alleles. Only the latter yielded white flowers because of the absence of a dominant red allele. Comparable experiments revealed similar percentages for other third-generation characters, as well as additional patterns of allelic dominance: the "rough seed" allele was dominant over the "smooth seed" allele, for example, and the "yellow seed" allele was dominant over the "green seed" allele. In essence, what Mendel's experiments revealed was: that for each variable character studied, an individual plant possessed a potential determinant (gene) from each parent; that one determinant masked the other when both were present; and that the two determinants were separated and passed on to offspring independently—they were retained as discrete particles or factors of inheritance.

Shortly after 1900, biologists came to appreciate that if genes are shuffled and reshuffled by random mating within a population of the sort studied by Mendel, the population will tend to undergo no net change. Its characteristics will, on the average, tend to remain the same. Chance fluctuations in gene frequencies can nonetheless yield some net change. In nature, however, natural selection yields more change. The raw materials of selection are mutations, and the most elemental mutations affect single genes. Genes are chemical structures, and because their mutations are therefore chemical changes, they are reversible. By the same token, a given kind of mutation can also occur more than once. Still, there has to be a first appearance for every allele, and the debut of any beneficial allele represents an important accidental step. Being accidental, most mutations are harmful. (A change that occurs by chance is not likely to improve a complex living system.)

Mutation continually expands the domain of variability of a population, giving evolution new boundaries in which to operate. Perhaps some day there will arise a new kind of pea seed unknown to Mendel—a seed not round or wrinkled, but ribbed. Perhaps for some reason this seed will be unusually fertile in poor soil and will be of great advantage to gardeners. If so, it will be selected for artificiality. In nature, a similar seed would, in the presence of poor soil conditions, be favored by natural selection and would expand the species' breadth of adaptation. If it cropped up in a small population, this kind of mutation might be associated with speciation; it might become fixed within a new kind of species.

Obviously mutation itself can disrupt the character of a population even before being "tested' by natural selection. A new kind of mutation is disruptive in a new direction. This fact quickly caught the attention of evolutionists after the turn of the century, when genetics was born. It was for this reason that the rediscovery of Mendel's work did not free Darwinism from its beleaguered state. On the contrary, with the new appreciation of genetic mutation, an over-reaction set in. Those, like William Bateson, who had previously opposed Darwin in emphasizing discontinuous evolution, found a new focal point: what came to be known as the macromutation or the mutation of great effect. Why look further than mutation, was the question. Why look beyond, to Darwinian selection? Strikingly unusual mutants could, in fact, be observed, especially in the plant world, and they seemed to represent an immediate mode of change that was both simpler and more tangible than Darwin's mechanism, which entailed the gradual accumulation of infinitesimal steps, each of which required at the least one minor mutation of the right kind.

Not only Bateson, but other prominent biologists, such as Hugo de Vries and T. H. Morgan, leapt upon macromutation as a panacea. They demoted natural selection to a minor sorting role. This viewpoint gained appeal in some quarters from a continued belief in Lord Kelvin's abbreviated estimate of the earth's age. Unlike natural selection, macromutation needed very little time, and, thus De Vries, in particular, rose to the stature of hero during the first decade of the century. In complementary fashion, Darwin's scheme of evolution declined to its all-time nadir of popularity. It became widely believed that species arose full blown by single macromutations. To many biologists it appeared that Darwinism was dead.

So disruptive was the turmoil created by the macromutationists that even wilder evolutionary schemes sprang up within the chasm left by the retreat of Darwinism. Some of these invoked neo-Lamarckian ideas or mysterious vital forces. One bizarre concept that was claimed to reflect fossil evidence was orthogenesis—literally "straight-line" evolution, or evolution impelled by unknown internal forces that were supposed to drive some species along linear

Darwinism Challenged, Darwinism Affirmed

paths of change, ultimately to extinction. This chaotic turmoil persisted into the 1930s, although after 1910 there were increasing attempts to bring Darwin's belief in slow, mechanistic process of selection back into dominance.

Such was the climate in which the Modern Synthesis of Evolution was born. The synthetic movement, which was nucleated by developments in the field of genetics during the 1920s and 1930s, brought evolutionary biology essentially full circle—from the anti-Darwinian posture of the first decade of the century back to Darwinism, but to a new, more sophisticated version of Darwinism. The Modern Synthesis represents the point of departure for many of the arguments that will appear in the chapters that follow. Accordingly, several facets of the movement deserve special scrutiny.

Shortly after the turn of the century, it was learned that genes are linearly arranged on chromosomes, elongate bodies that lie within the nuclei of cells. The 1920s saw the birth of modern population genetics, a field that treats changes in gene frequencies mathematically. Mutation, of course, found a home in the new evolutionary genetics, but in a scaled down fashion—not as the macromutation that had been believed capable of instantly creating a new species, but as a process continually generating variability randomly and by small degrees. Geneticists came to stress the detrimental nature of most mutations. From this emphasis it seemed to follow that the more pronounced was the change wrought by a mutation, the less likely it was that the mutation would be valuable. Only occasional, small mutations were considered likely to improve adaptation. The concept of the macromutation was discarded.

The particular aspect of the Modern Synthesis that I wish to single out is the restored Darwinian notion that evolution proceeds slowly—that the testing of mutations is a process of infinitesimal steps. A primary reason for this gradualistic orientation of the Modern Synthesis was unquestionably that the new genetics was revealing evolution before the very eyes of experimenters. Theodosius Dobzhansky, who by mid-century became the leading experimental geneticist, was a prime mover here. After migrating from Russia to America, he showed that natural populations of the fruit fly *Drosophila* undergo seasonal genetic alterations that represent selection by the changing environment. Furthermore, he could actually arrange for similar evolution to occur in the laboratory—evolution that represented adaptive response to artificial external conditions. In 1947, he wrote:

> Controlled experiments can now take the place of speculation . . . The mechanics of natural selection in concrete cases can be studied. Hence the genesis of adaptation, which is possibly the central problem of biology, now lies within the reach of the experimental method.[12]

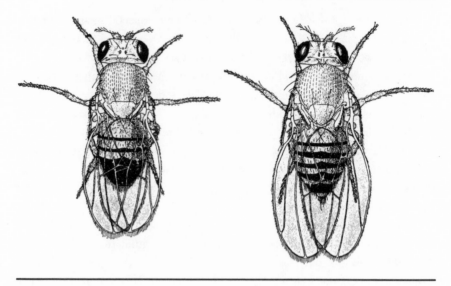

FIGURE 4.3

Drosophila melanogaster, the famous fruit fly that geneticists have studied extensively in the laboratory. The male is shown on the left and the female, on the right.

A new world had suddenly opened to the evolutionist. As we have seen, evolution, though generally accepted, had fallen from favor as a field of inquiry in the late 1800s. Now Dobzhansky and others claimed to be bridging the gap between fact and speculation. They were elevating evolutionary science from its historical, conjectural level of operation to the status of a rigorous experimental science resembling physics and chemistry. Furthermore, with the development of quantitative population genetics, biology, like the physical sciences, was becoming grounded in mathematical theory.

The Modern Synthesis was reflected in and, in part, spawned by two major books. Both deal with its gradualistic theme. One was Dobzhansky's *Genetics and the Origin of Species,*[13] the very title of which, borrowing from the name of Darwin's book, brought Darwinism again to the fore, but in light of the new approach to inheritance. Although Dobzhansky was an experimentalist, in discussing where evolution occurs most effectively, he turned to the theoretical work of others instead of relying upon his own experiments or observations. In particular, he leaned upon the mathematical analyses of Sewall Wright. Wright had made an important contribution by showing that small populations were subject to chance changes in gene frequencies (accidental evolution). With mathematical rigor he confirmed what we all know intuitively: when we are dealing with very small numbers, things become unpredictable; accidents loom large. To use the common phrase defined in chapter 1, genetic drift is most important in small populations. In large populations, on the other hand, accidents tend to average out. Dobzhansky promulgated Wright's resulting argument that because of the elevated importance of chance factors, natural selection will tend not to work effectively in small populations. An emphasis on natural selection as the chief mechanism of change seemed to require an emphasis also on large populations.

There was another side to Wright's coin that Dobzhansky also presented in a favorable light. Wright argued that natural selection will operate most effectively in large populations divided into many partly isolated subpopulations. The idea is that this kind of population structure promotes variability upon which natural selection can operate. A favorable mutation that happens to crop up and then to become fixed by selection within one of the subpopulations may turn out to have value to the species as a whole. Partial isolation fosters numerous opportunities for such occurrences, but then some interbreeding among subpopulations (incomplete or impermanent isolation) permits generally useful but locally established features ultimately to spread throughout the entire species.

Thus, Dobzhansky argued that large species are readily transformed. This notion was entirely compatible with Darwin's belief that large, well-established species are constantly evolving at significant rates. Darwin's gradualistic view was thus perpetuated. Like Darwin, Dobzhansky largely ignored the possibility that most significant evolutionary changes might take place rapidly in small populations diverging to become new species. Also like Darwin, he had almost nothing to say even about the manner in which species multiply. In an article published in 1972, Dobzhansky did write about speciation, but he described the gradual divergence of a new species from its ancestor as the "usual, and by now orthodox, view."[14]

The second seminal book of the Modern Synthesis gave the movement its name. This was *Evolution: The Modern Synthesis*,[15] by Julian Huxley, grandson of Thomas Henry Huxley, "Darwin's Bulldog." Huxley reiterated Wright's model for population size, with strong agreement:

> It is furthermore obvious that only abundant and widespread species will be of any service in tracing the detailed course of past evolution. . . . In the first place, abundant species will have a larger reservoir of inheritable variation, both actual and potential. This can be deduced on theoretical lines from what we know of mutation (Wright, 1932). In addition, it has been demonstrated as a fact . . . in

several places. Darwin, on the basis of qualitative inspection, asserted that it was so. . . . This will obviously confer on abundant species a greater evolutionary plasticity, a higher potency of adaptive change.

Rare species, on the other hand, will not only possess less evolutionary adaptability, but will, as Sewall Wright (1932) has emphasized, be prone to have useless or even deleterious mutations become accidentally fixed. . . .[16]

Huxley devoted more attention than Dobzhansky to the multiplication of species, but still saw speciation as contributing nothing to the pace of evolution: "A large fraction of it," he wrote, "is in a sense an accident, a biological luxury, without bearing upon the major and continuing trends of the evolutionary process."[17] Here again, the theme of gradualism prevailed.

I will argue that there is a fallacy in these long-prevailing arguments that large, patchily distributed species form the locus of the most rapid evolution. In fact, it is in such populations that evolution is likely to proceed most slowly; they are strait-jacketed by their very diversity and complex geographic distribution. Instead, to find rapid evolution we must look to very small, well-mixed, inbreeding populations—ones that occupy restricted, homogeneous, and often novel habitats where severe selection pressure can work quickly and effectively.

The gradualism of the Modern Synthesis was given encouragement by the removal of the two obstacles that had frustrated Darwin: the alleged brevity of Earth history and the concept of blending inheritance. Armed with seemingly limitless time after the discovery of radioactivity and particulate inheritance, the new geneticists deemed it reasonable to extrapolate the results of their exciting laboratory experiments to selection within well-established species in the natural world.

In part, too, the new gradualism represented reaction to the excesses of those evolutionists who speculated broadly during the early part of the century. Many who united under the banner of the Modern Synthesis had been confronted in their early professional lives with evolutionary schemes that, though unscientific and embarrassing to the profession, were given some credence because neo-Darwinism was not well enough developed to stand its ground. When a more fully-formed theory of selection, built upon Darwinian gradualism and modern genetics, began to take shape, it is no wonder that bright scientists embraced it almost unanimously.

The momentum gathered by the Modern Synthesis was evident in the events of the centennial year of the *Origin of Species*. In 1959, a number of biographies of Darwin were issued, and many symposia were held and published in celebration. While the many valid aspects of Darwin's work were supported, so

Darwinism Challenged, Darwinism Affirmed

was his gradualism. What is unfortunate is that, in a political sense, there was little room for dissent. This is not to say that there were no nay-sayers. Throughout the 1940s, for example, the plant biogeographer J. C. Willis and the geneticist Richard Goldschmidt argued for the sudden appearance of species by macromutation. Because of his prominence as an experimentalist, Goldschmidt was too conspicuous to be ignored and he, in particular, was ostracized by the evolutionary community.

While Willis and Goldschmidt were extreme in their views, they made major contributions by pointing science in potentially useful directions. I will argue that they erred no farther in one direction, however, than did typical contributors to the Modern Synthesis in the other. Furthermore, a small number of biologists adopted the punctuational view that most major evolutionary transitions occur in small populations. These were generally individuals, such as Ernst Mayr and Verne Grant, who concerned themselves with species as units of evolution. Because their views were not extreme, these men were able to contribute to the Modern Synthesis. Significantly, however, their arguments for rapidly divergent evolution in small populations were for many years ignored.

The Modern Synthesis was perhaps not so much a true synthesis as it was a victory for gradualistic genetics. The evolutionary discipline least happily accommodated was the one about which I have said least but in the following chapter will say most: paleontology or, as it is commonly called today, paleobiology. The known fossil record is not, and never has been, in accord with gradualism. What is remarkable is that, through a variety of historical circumstances, even the history of opposition has been obscured. Few modern paleontologists seem to have recognized that in the past century, as the biological historian William Coleman has recently written, "The majority of paleontologists felt their evidence simply contradicted Darwin's stress on minute, slow, and cumulative changes leading to species transformation."[18] In the next chapter I will describe not only what the fossils have to say, but why their story has been suppressed.

CHAPTER

5

The Fossil Record: Our Window on the Past

I F THE RECORD of the rocks had never been, if the stones had remained closed, if the dead bones had never spoken, still man would have wondered."[1] With these words Loren Eiseley paid indirect homage to the fossil record. Without our partial catalog of ancient life, we would indeed have wondered, but would we have believed? It is doubtful whether, in the absence of fossils, the idea of evolution would represent anything more than an outrageous hypothesis. Certainly it would still arouse skepticism.

The fossil record, and only the fossil record, provides direct evidence of major sequential changes in the Earth's biota. On a finer scale, paleontology is our only direct source of information about the course of evolution within segments of phylogeny: through it we can, with varying degrees of confidence, identify particular ancestral and descendant groups.

Another unique offering of the rock record is a time scale for evolution. I see no way other than by studying geological time—the third dimension of the distribution of life on Earth—that we can test the gradualistic view. Biologists cannot observe nature fully, nor can they mimic her perfectly in their laboratories. Limited by their own lifetimes, they occupy such a narrow sliver of Earth history that their experiments and observations cannot be extrapolated to the necessary spans of time. One problem is that of scaling up the observations or experiments of days or months in order to assess events that in nature occupied millions of years. Another difficulty is our inability to witness, at first hand, events that are rare even on a geological scale of time. Although some species have undoubtedly originated during our recorded history, no human has ever seen a new species form in nature.

Before we examine some of the things that fossils now reveal about the tempo of evolution, we should recall Darwin's gloomy appraisal of the fossil record. He acknowledged in the *Origin* (p. 299):

> Geological research, though it has added numerous species to existing and extinct genera, and has made the intervals between some few groups less wide than they otherwise would have been, yet has done scarcely anything in breaking down the distinction between species, by connecting them together by numerous fine, intermediate varieties; *and this not having been effected is probably the gravest and most obvious of all the many objections which may be urged against my views.* [italics added]

Darwin's escape, of course, lay in his condemnation of the record. To this end he dedicated an entire chapter, entitled "On the Imperfection of the Geological Record." The depth of Darwin's concern about the failure of fossils to support his grand scheme is expressed by his bald statement in the *Origin* (p. 342): "He who rejects these views on the [incomplete] nature of the geological record, will rightly reject my whole theory." In fact, we need not go so far. As I will attempt to show in the chapters that follow, only modification of the theory is required, not utter rejection.

It should also be recalled that when Darwin penned the *Origin,* rather little was known about the fossil record—so little that Lyell, though recognizing that species come and go, was still able to resist the idea of large-scale biotic change. Thus, the first edition of the *Origin* was almost without helpful paleontological examples. New discoveries during the 1860s and 1870s permitted Darwin to include more supportive fossil data in later editions. It is nonetheless important to understand that even these data failed to provide the ultimate evidence that Darwin sought—evidence of gradual transitions from genus to genus or family to family. What did come to light were sequences that could, with a little imagination, be connected to form approximate pathways of evolution.

The greatest success in recognizing general pathways of evolutionary descent

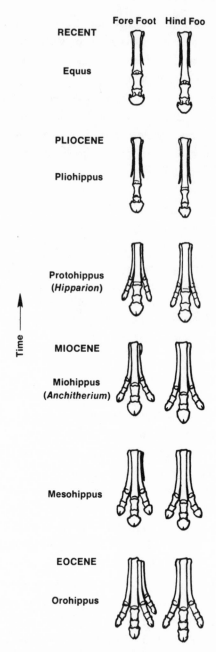

Fore Foot   Hind Foo

RECENT

Equus

PLIOCENE

Pliohippus

Protohippus
(*Hipparion*)

MIOCENE

Miohippus
(*Anchitherium*)

Time ⟶

Mesohippus

EOCENE

Orohippus

FIGURE 5.1

General stages in the evolution of the modern horse, as recognized by O. C. Marsh in 1879.

came in the study of relatively youthful fossils—mammal remains of the Cenozoic Era. The Frenchman Albert Gaudry and the Russian V. O. Kovalevskii were among the first paleontologists to make considerable headway here, and among their discoveries were skeletons that seemed to represent several stages in the development of the modern horse. The importance of the fossil finds of the years following publication of the *Origin* in 1859 becomes obvious when one considers that in the 1850s only distinctive fossil faunas of various ages were recognized. No well-delineated succession of fossils for any one family of animals was obtained. In addition to the general sequences, there appeared a few fossils of the sort commonly hailed as "missing links." The most notable of these was *Archaeopteryx,* the famous German discovery of 1861. This creature was about the size of a crow. T. H. Huxley correctly showed it to be intermediate between small dinosaurs and birds. It had the skeleton of a dinosaur and the feathers of a bird. There is now a widespread belief that birds evolved from dinosaurs. All this patchy evidence was tantalizing. It pointed toward descent, but what still failed to materialize were the *gradational* changes predicted by gradualism—slow, continuous, species-to-species transitions connecting genera or families. These would have confirmed Darwin's gradualistic view of evolution. The result was an ambivalent reaction of the paleontological community towards Darwin's scheme. As required by their general sequences of fossils, paleontologists accepted the idea of large-scale evolution, but they tended to denigrate natural selection as the prevailing mechanism. This negativity followed from the fact that Darwin had wedded the concept of natural selection to gradual change, for which paleontologists found little evidence. Their protestations were not well received, however. Aided by Lyell's *ex cathedra* blessing, Darwin's deprecation of the fossil record took root. Almost officially, the mission of paleontology became restricted. Fossils were thought to be good for little more than roughing out broad patterns of certain portions of the tree of life. This amounted to little more than collecting, describing, classifying, and conjecturing.

Thus, paleontology joined the conceptually weak and speculative areas of biology that late in the nineteenth century engendered disillusionment with evolutionary research. Joseph Leidy, one of the most prominent paleontologists of the day, introduced a monograph on fossil mammals of Nebraska with this statement:

> The present work is intended as a record of facts in paleontology, as the authors have been able to view them; a contribution to the great inventory of nature. No attempt has been made at generalization or theories which might attract the momentary attention and admiration of the scientific community.[2]

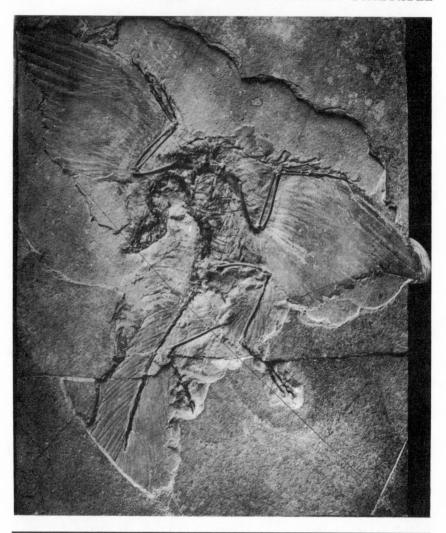

FIGURE 5.2

Archaeopteryx, an extinct Jurassic animal intermediate in form between dinosaurs and birds.

Those few paleontologists who were inspired to engage in evolutionary theorizing tended to opt for fanciful schemes. The Americans, Alpheus Hyatt and Edward Drinker Cope, emphasized the Lamarckian idea that characters acquired during life are heritable, and these "neo-Lamarckians" further argued that it was not natural selection but the interaction of the individual with its environment that effected change: in each generation, a giraffe stretched its neck to reach food on high branches, and this induced growth of a slightly longer neck, which was passed on to offspring.

During the early part of the twentieth century, when evolutionary theory was in turmoil, paleontologists continued along the same path. Strangely, despite Darwin's work, many remained typologists, naming new species from sparse fossil remains (often a single specimen) that differed only slightly from other material. Henry Fairfield Osborn, who as the powerful director of New York's American Museum of Natural History dominated vertebrate paleontology early in the century, was guilty of this kind of taxonomy. He also issued numerous arbitrary assertions about the mechanism of evolution. These are difficult to decipher, but were in part non-Darwinian.

When the Modern Synthesis emerged at about the time of the Second World War, paleontologists were not among its primary founders. Still, as I will describe more fully later in this chapter, they joined in the new movement to the point that by 1959, the centennial year of the *Origin of Species*, nearly all students of the fossil record accepted Darwinian gradualism. At least it was a rare professional who was willing to put dissenting views into print.

In the biological sphere, the modern punctuational view of evolution was first suggested by Ernest Mayr, formerly of Columbia University and now of Harvard. As early as 1942, Mayr[3] argued that some living genera of birds have evolved rapidly, each as a small, marginal population diverging from its ancestral species. In 1954, Mayr[4] generalized further, pointing out that many important evolutionary transitions have taken place rapidly in small, local populations. He further noted that this mode of evolution could explain the seemingly sudden appearance in the fossil record of many "evolutionary novelties." The point here is that if the transition was typically rapid and the population small and localized, fossil evidence of the event would never be found. The other aspect of this argument is that the general failure of the record to display gradual transitions from one major group to another did not reflect a poor record for large, well-established species, but the slow evolution of such species: full-fledged species are not the entities that undergo the majority of major evolutionary changes.

Mayr's punctuational arguments flowed directly from his studies of the manner in which species multiply. Distinctive species of birds on islands, for example, seemed to demand a theory for rapid origination. Mayr was the chief architect of the view that most species evolve from geographically isolated populations. Although Mayr reasserted his punctuational views in the widely heralded book *Animal Species and Evolution*[5] (1963) and became recognized as one of the leading evolutionists of the century, little attention was paid to the punctuational elements of his work until the 1970s. This paradox was partly the result of the diffuse but ever-present counterpressure supplied by the field of genetics, in which Mayr is not a specialist. The gradualistic march of the geneticists had gathered too much momentum to be diverted by peripheral activities.

In the early 1970s, two paleontologists, Niles Eldredge of the American Museum of Natural History and Stephen Jay Gould of Harvard University, urged that Mayr's views be given credence. They first applied the terms "punctuational" and "gradualistic" to the alternative views of evolution. Their writings drew considerable attention to the basic question of alternative views.[6,7] Unfortunately, an exaggerated polarization developed. Some workers assumed that the punctuational view virtually denied evolution within established species; conversely, they assumed that the gradualistic view saw speciation as almost never being rapidly divergent. Because it then seemed evident that the truth lay on middle ground, the controversy seemed meaningless or misleading. As a result of this misunderstanding, I have suggested that we erect unambiguous and nonextreme models: the *gradualistic model,* representing the traditional view, holds that most evolutionary change in the history of life has taken place within fully established species, while the *punctuational model* asserts that most change is associated with speciation that involves small populations. If about half of all evolution were attributable to small, diverging populations and about half to large, established species, it would be meaningless to contemplate these models, but I believe otherwise. In the following paragraphs, I will attempt to show how most genera, families, and still larger branches of the tree of life have developed mainly by way of one or more steps of rapid evolution associated with branching (speciation).

During the belated outburst of debate over the punctuational view in the early 1970s, it was asserted that the fossil record could not settle the issue. One could not choose whether to interpret the lack of fossil evidence for gradualism as a failing of the record or as a failing of gradualism. This assessment is overly pessimistic. The fossil record is very bad, but it is also very good. To take advantage of its strengths, we must look in the right places and ask appropriate questions.

To appreciate how bad the record can be, we need look no further than the largest class of living animals, the insects. More than a million species of insects inhabit the world today. This means that during the entire Cenozoic Era (the past sixty-five million years) tens of millions of insect species have existed. Of these, only a minute fraction have been discovered as fossils. Insects are small, fragile animals that are almost perfectly preserved only if trapped in amber, which is fossilized tree resin. Amber, of course, is so rare as to constitute a semiprecious stone. It is easy to see why the fossil record of insects is relatively poor. The record of insects nonetheless illustrates how careful scrutiny often yields crucial data. This record offers a powerful line of evidence that represents the kind of argument that I will stress in this chapter. The evidence comes in the form of beetle genitalia! Beetles are very important creatures, forming the largest order of animals on Earth, with more than 300 thousand species. The great population geneticist J. B. S. Haldane, when queried by a prominent clergyman as to what traits of the Creator were evidenced by life on Earth, responded "an inordinate fondness for beetles." Living beetles are assigned to species according to the shapes of their external skeletons, especially the genital portions. Beetle skeletons are stubbornly durable, and they are well preserved in geological settings that remain damp and protected from atmospheric oxygen. G. R. Coope of the University of Birmingham has assembled all available data and found that beetles of the Ice Age are almost identical to living species. This means that no lineage that is known to have passed through a million years or two (and the same number of generations) has undergone appreciable evolution!

FIGURE 5.3
Well-preserved, long-snouted weevil from Alaskan Ice Age deposits older than 1.5 million years. This insect belongs to the Coleoptera, the order of insects containing beetles, which produce resilient fossil remains.

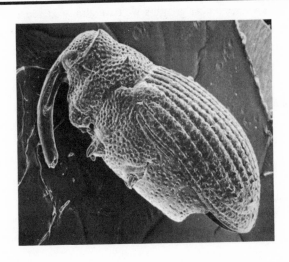

Insects are not the least easily fossilized of animals. Many groups of soft-bodied animals and plants are unknown in a fossil state. For animals and plants with durable skeletons or other resistant structures, the story is quite different, however. Vertebrate animals (animals with backbones) are commonly preserved, although only for Cenozoic mammals is the record good enough to allow us to test many evolutionary hypotheses. This is no problem if the hypotheses are of a general nature and the results can be extended to other forms of life. A variety of invertebrate animals (animals without backbones) also have readily preservable skeletal structures. Most fossil invertebrates have lived in the oceans, which represent vast depositional basins where sediments (primarily sand and mud) accumulate and entomb some skeletal remains. Many limestones are actually formed primarily of bits and pieces of limey invertebrate skeletons. It has been estimated that about 30 percent of all marine invertebrate species of moderate body size can leave fossil records. The actual record is not this complete, nor, of course, is it fully studied. Some of the record is lost because layers of sediment are entirely destroyed by erosion. Others are buried and altered by intense heat and pressure beyond recognition—or beyond the point where their fossils can be recognized. Nonetheless, some groups of marine invertebrates are well fossilized. One expert has estimated that for sea urchins, roughly one-half of all species that existed during the Cenozoic Era have actually been discovered in a fossil state and described.

To a considerable degree, the secret of science is the secret of asking the right questions. Endless queries can be addressed to nature. The key to success lies in choosing from the myriads of possibilities those particular questions that are both important and answerable. Many of the best questions are almost rhetorical, which is to say ones whose answers we have almost at the asking. Thus, we are often wise to pass over questions that are perplexing, though important, to ones whose answers jump out at us. The spectrum of appropriate questions shifts through time, as data, techniques, and theories expand. Today we can interrogate the fossil record in ways that were impossible during the last century. Our enlarged opportunities result not only from our increased knowledge of the fossil record, but also from our possession of an absolute scale of geological time.

The relative time scale was established primarily during the last century. It did not develop logically or in chronological order. Like Topsy, it just grew. As the pieces of Earth history fell into place, the three great eras represented by highly fossiliferous rocks—the Paleozoic, Mesozoic, and Cenozoic—came to be divided into formal "periods," and these into "epochs." To some degree, the

The Fossil Record: Our Window on the Past

positions of present boundaries between intervals reflect historical accident. Not all are meaningful divisions. Also, the bodies of rock representing the time units are not perfectly delimited throughout the world. In Western Maryland, I direct my students' attention to the seemingly plain face of a particular limestone quarry and tell them that somewhere in that rock wall lies the boundary between the Silurian System and the Devonian System. The level of the boundary is estimated from the study of fossils harbored within the limestone, but the Silurian and Devonian are intervals defined across the Atlantic in Britain, where the fossils are not exactly the same. The truth is that we do not know exactly where to draw a boundary line on our quarry wall in the Appalachians.

Shortly after the turn of the century, the discovery of radioactivity not only provided a heat source to account for sustained high temperatures for the Earth, it also gave us a series of clocks for evaluating geological time. Any radioactive element (isotope) decays at a particular average rate. Knowing this rate and the amounts of parent and daughter isotopes in a rock, we can calculate the time when the mineral containing the parent isotope was formed. This approximates the age of the rock. This technique is ruled out for most sedimentary rocks, however, because their component minerals either are not radioactive or are present as old particles washed into a depositional basin that is much younger than they. Even so, ages can sometimes be estimated for undatable sedimentary rocks from dates obtained for intimately associated crystalline rocks—rocks such as those produced by lava that flowed into a sedimentary basin or granites formed by the intrusion of magma into fresh sediments.

It is not widely appreciated that radiometric dating of the sort I have just described provides only an approximation. We must contend with errors of measurement and imperfect preservation of parent and daughter isotopes. As it turns out, fossil data are not only more often available, they also commonly provide better accuracy in relating a body of rock to the standard geological time scale. What radiometric methods have done, almost uniquely, is to convert the relative time scale to an absolute time scale. Thus, "the Cambrian preceded the Ordovician, and the Ordovician preceded the Silurian. . . ." becomes "the Cambrian began about 570 million years ago, and the Ordovician began about 500 million years ago. . . ."

On a finer scale, we can use absolute dates to estimate rates of evolution. We might, for example, measure the percentage of evolutionary change in body size or shape per million years for the transition from one form to another. We can also estimate how long the fossil entities that we call species have survived in geological time. In the nineteenth century, no such calculations were possible. One of the initial decisions in asking evolutionary questions of the fossils relates to the choice of an interval for study. The place where we can most effectively make use of radiometric dating is in rocks of the Cenozoic Era—the Age of Mammals. Here the errors of dating are relatively small and the fossil record is of high quality. Shells and bones have not been as badly damaged by natural chemical agents as in older deposits, and being at the top of the heap, Cenozoic sediments are also widely exposed around the world. In addition, they remain largely unlithified (soft). It is much easier to recover a fossil by plucking or sieving it from soft sand or mud than by attempting to break it from solid rock (where the fossil may, in fact, be difficult to find in the first place). Another point in favor of Cenozoic fossils is their relatively close relationship to living creatures. Just how this works to our advantage will become apparent in some of the discussion that follows.

The study of fossils has been disparaged for lacking an experimental approach. As we have seen, paleontology in the late nineteenth century was indicted for its speculative nature. The implication was that its fundamentally historical nature required it to proceed by conjecture. This was, in fact, a false charge. Unbridled speculation is not inherent to the study of fossils, but happened to characterize it in the decades immediately following publication of the *Origin of Species.* A. Conan Doyle's Sherlock Holmes often operated in the manner of the historical scientist, reconstructing past events from what scraps of evidence were available after the fact. His mode of choosing particular questions to ask and facts to gather from among an infinite number of possibilities was closely analogous to the procedure followed for effective sleuthing in the fossil record. Holmes' deductive reasoning was comparable.

It is true that we can seldom test a paleontological hypothesis with a live experiment. What we can do is to perform "thought experiments." We can erect hypotheses, deduce their consequences, and then see whether these consequences are borne out by fossil data. Because we are seldom able to conduct the experiments ourselves, we seek out critical experiments that nature has performed for us. Sometimes they have occupied millions of years and entailed large and enormously complex systems. Nature's laboratory is unmatched in scale and sophistication.

Let me illustrate with an experiment that I find simple but compelling. Appropriately, it serves to test the punctuational model against the gradualistic model. I will withhold the name of the test for the moment, because it anticipates the outcome. The test concerns particular portions of the tree of life, namely narrow branches, or ones with few divisions. A branch of this type might represent an entire family or order of animals or plants, but one that at no time has included more than a very few species—ideally two or three.

In the gradualistic scheme, because lineages tend to evolve at a moderate pace, narrow branches, like all others, should tend to undergo moderately rapid transformation. But what evolutionary fate would the punctuational model predict for such narrow branches? This model relies on speciation (branching) for most evolutionary change. If branching is where most of the action is, then narrow segments of phylogeny—ones that have not included much branching—should tend to display little net change.

Here then we have a test. We can single out for examination narrow branches of phylogeny—small but persistent families or orders—and we can then see how much net change they undergo. In other words, is the original genus of the group eventually succeeded by others of quite different form (gradualistic model), or are the changes minor, so that even over long stretches of time no new genera, or only slightly different ones, emerge (punctuational model).

We must bear in mind that in the fossil record we have available only bits and pieces of phylogeny. If we had a complete picture, we could read the pattern (gradualistic or punctuational) directly. We would need no test of the sort I have outlined. The test is useful because all that it requires are two things. First, we must know the kinds of animals or plants forming the base of the branch and the younger ones forming the tip. Second, we must have a large enough number of sequential samples in between to indicate that we are dealing with a coherent natural group (a real branch) and to show that the group never included many lineages.

As a first example, we can consider the bowfin fishes, known as the family Amiidae. These are large, exclusively North American freshwater animals, which have an excellent fossil record in sedimentary deposits that represent several epochs of the Cretaceous (the final period of the Mesozoic Era) and every epoch of the ensuing Cenozoic Era. No more than two bowfin species are known to have existed at any one time. Thus, all the criteria for the test are met. What has happened to the bowfin fishes during their long history of more than one hundred million years? Next to nothing! During the latter part of the Cretaceous, bowfins became slightly more elongate, but during the entire sixty-five million years of the Cenozoic Era, they evolved in only trivial ways. Two new species are recognized, but these differ from their Late Cretaceous ancestors only in subtle features that represent no basic shift of adaptation. The bowfins of seventy or eighty million years ago must have lived very much as their lake-

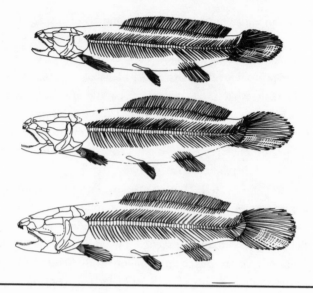

FIGURE 5.4

Close resemblance among a sixty-five-million-year-old bowfin fish of the Early Cenozoic (below), a mid-Cenozoic bowfin (center), and a modern bowfin (above). The latter is a classic living fossil.

dwelling descendants do today. Thus, the punctuational view is favored.

The lungfishes represent another example. These animals are well equipped to withstand droughts that occur where they live in South America, Africa, and Australia. Lungfishes breathe air, and the South American and African forms can also burrow into mud in order to sit out dry intervals. The evolutionary history of the lungfishes differs from that of the bowfins in that it includes an early history of rapid diversification. The lungfishes evolved at a high rate early in their history, when they also speciated rapidly. Then, beginning about 300 million years ago, they declined to a small number of lineages. At this low diversity they persisted to the present day. Meanwhile they evolved only modestly. Again, the punctuational model is favored.

When we conduct this test with other groups, we find much the same pattern. Families and orders that have persisted for long intervals by way of a small number of lineages have retained their initial anatomical shapes. Among the other small groups of animals exhibiting the characteristic pattern of evolutionary stagnation are sturgeon fishes (known for their caviar), garpikes, snapping turtles, alligators, tapirs, and aardvarks.

The living representatives of all the small groups of animals I have just

The Fossil Record: Our Window on the Past

described are what are commonly referred to as "living fossils." Thus, I have called the test just outlined the test of living fossils. Use of this label should not be taken to imply that only groups known in advance to be living fossils are considered. Such a procedure would represent a fatal bias because it would automatically exclude from consideration any narrow branches of phylogeny that do display major changes. Rather, I have tried to include in the test all narrow branches for which we have a fairly good fossil record. The critical point is this: I have yet to uncover one of these that displays marked evolutionary change.

Living fossils have represented a thorny puzzle in the traditional, gradualistic scheme of evolution. If natural selection is constantly reshaping species in significant ways, why have some species been almost immune to the process? Darwin, who seems to have coined the phrase "living fossils" in the first edition of the *Origin* (p. 107) suggested that "they have endured to the present day, from having inhabited a confined area, and from having thus been exposed to less severe competition." On the contrary, we can now see that many species of living fossils are not narrowly distributed. Delamare-Deboutteville and Botosanéanu,[8] who recently published a book on living fossils, described them as creatures "that have stopped participating in the great adventure of life," being confined by their narrow adaptations. Exactly the opposite explanation has been offered by George Gaylord Simpson, who has considered living fossils to have stagnated because of their unusually broad adaptations; according to his view, a typical living fossil species has tolerated such a wide variety of conditions that it has not been subjected to strong, specific pressures of natural selection.

These conflicting conjectures on ecological breadth illustrate the dilemma in which gradualism has been trapped by living fossils. In truth, some living fossils are narrowly distributed and others are broadly distributed. Some are narrowly adapted and others are broadly adapted. Many, like the American alligator and snapping turtle, are quite abundant, or were prior to human interference. Living fossils share no obvious adaptive feature that can explain why natural selection should have largely ignored them for millions of years while working enormous changes on other well-established forms of life.

The punctuational model provides us with the explanation. To begin with, the fossil record shows that many small groups of animals and plants survive for short intervals, but a few survive for long periods of time. The punctuational model then predicts that even the persistent small groups will tend to exhibit little evolution because, however long they may have existed, they have undergone little speciation—and speciation is where most evolution occurs. In other words, the punctuational model, rather than leaving us stymied, predicts the existence of living fossils. They are the inevitable end products of long, narrow branches of the tree of life.

We can put the gradualistic model to another test that has broader application and penetrates to the very heart of large-scale evolution. This test relates to our knowledge that distinctive groups of animals and plants have come into being during particular intervals of time. We can often establish a maximum age for a family, order, or class by noting when its ancestors originated. The test is to determine, in this context, how rapidly the established species of the group evolved. For example, the fossil record indicates that whales evolved from small terrestrial mammals during, at most, twelve million years. Have fully-established species of mammals generally evolved rapidly enough for their transformation to account for this rapid origin of whales, as gradualism would demand? Or are well-established species modified so slowly that we must invoke rapid, stepwise branching to achieve the rapid conversion of a small land animal to a large ocean-going swimmer?

The general phenomenon that I am addressing here is what we call adaptive radiation. Adaptive radiation is the rapid divergence of many new forms of life from some common ancestor. Adaptive radiation has accounted for most of the large-scale evolutionary changes in the history of life. The most spectacular adaptive radiation of all time, at least for multicellular life, occurred near the beginning of the Cambrian Period. Simply in recognizing this example of adaptive radiation, I am presenting a conclusion that was beyond reach until quite recently, and was certainly so in Darwin's day. For many years, the fossil record of advanced life seemed to appear suddenly, in rocks more than half a billion years old, more or less at what we recognize as the base of the Cambrian, but it was not clear that this resulted from sudden evolution. What was clear early on was that this most obvious biotic discontinuity of the entire geological column deserved special recognition. It became formally designated as the boundary between the two most fundamental divisions of geological time: the Cryptozoic Eon, or "interval of hidden life," and the Phanerozoic Eon, or "interval of well displayed life." Thanks to radiometric dating, we now know that the Cryptozoic (commonly called the Precambrian) represents some four billion years, while the Phanerozoic has thus far lasted less than 600 million. In other words, for most of its history, our planet seems to have been barren of advanced life. Near the base of the Cambrian we find the first shelled animals—trilobite arthropods, snails, brachiopods, and a variety of less common, extinct, invertebrate marine animals.

The Fossil Record: Our Window on the Past

The relatively sudden appearance of the diverse Cambrian faunas was well known when Darwin wrote the *Origin* (although what we now call the Cambrian then formed part of the Silurian System). How could Darwin's gradualistic evolution be squared with the sudden appearance in the fossil record of varied forms of higher life? His only recourse was to conjure up a long, hidden interval during which higher organisms evolved. Perhaps sediments representing this interval lay hidden beneath modern oceans, or perhaps they had been buried and altered beyond recognition. Despite offering these conjectures in the *Origin* (p. 308), Darwin confessed, "The case at present must remain inexplicable; and may be truly urged against the views here entertained."

Among the most exciting fossil discoveries of recent years have been those made in rocks of Early Cambrian and latest Precambrian Age. Dating of such old rocks is imprecise, so that throughout most of the world the lowermost boundary of the Cambrian System can be determined in only an approximate way. Even so, remarkable patterns of fossil occurrences have emerged. Careful searching has turned up so-called trace fossils—tracks, trails, and burrows made by soft-bodied animals crawling on and through sand and mud. Trace fossils are formed on and within the sea floor today, and they are common in Phanerozoic rocks. Some fossil examples are recognized as the markings of particular kinds of animals. It can now be shown that trace fossils make their first appearance in the record as simple tubelike structures of very late Precambrian Age—in rocks older than those harboring the first faunas of shelled animals. What is particularly striking is that as one follows the trace fossil faunas up into the Cambrian, they display increases in variety and complexity of form. The simple tubes give way to branched tubes and to markings made by many-legged animals rather than by simpler wormlike forms.

In many parts of the world, in rocks of about the same age as those containing the earliest trace fossils, other markings of soft-bodied animals can also be found. These are imprints that reveal the shapes of several kinds of creatures. Many represent jellyfishes. One kind seems intermediate between a segmented worm and an arthropod. Others puzzle us with their unfamiliar shapes. The imprints and trace fossils tell of a strange, primitive world of fleshy animals, a world in which shells and skeletons were weak or absent. The fossil evidence also paints a primordial seascape in which large, sophisticated predators were lacking. Several kinds of animals (but not many by modern standards) were thus permitted, without benefit of armor, to crawl on or through the sea floor, float above it, or attach to it by stalk or stem.

Above the rock layers on which the curious imprints of fleshy animals are embossed and through which the earliest burrow tubes pass, we find the first

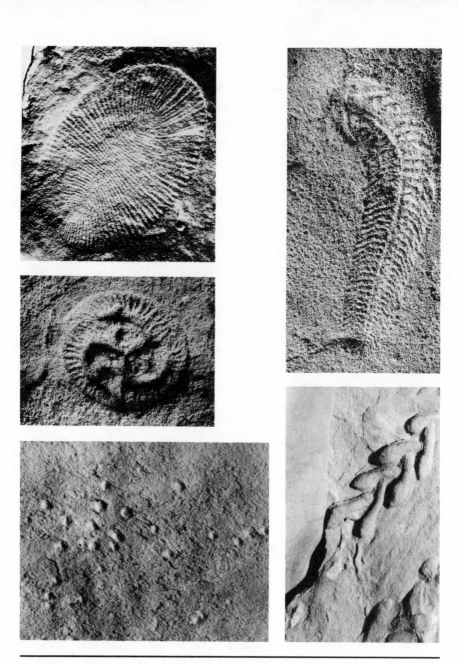

FIGURE 5.5

Fossils of the latest Precambrian (more than 600 million years old) representing very early multicellular life of the sea floor. Upper right: Impression of an animal perhaps intermediate between a worm and an arthropod. Center: A problematical form possibly related to a starfish. Lower left: Sediment fillings of tubes that are among the oldest and simplest trace fossils. Lower right: Sediment fillings of a slightly younger and more complex feeding track on the sea floor.

shelly fossils. It was long thought that the earliest of these were the trilobites—the primitive shelled arthropods so prized by hunters of fossils. During the past few years, however, discoveries in many regions of the globe have revealed skeletal faunas slightly older than the oldest trilobites. These consist of minute marine forms, including members of familiar groups like sponges and snails, but comprising for the most part strange tiny creatures, some of which seem unrelated to any other known living or fossil animal. It is estimated that this so-called Tommotian shelly fauna, along with some persistent soft-bodied creatures, populated the world's seas for between fifteen and thirty million years of the earliest Cambrian—a brief interval, representing less than 5 percent of all Phanerozoic time. These groups gave way to the trilobite-dominated faunas that reigned for the remaining sixty million years or so of the Cambrian Period. These, in turn, were succeeded by the more varied and more modern shelled marine faunas of the Ordovician Period. Shells are useful for tissue support, but they also serve to ward off predators. Undoubtedly the multiplication of shelled groups during the Cambrian was in part a response to the evolution of increasingly advanced marine carnivores. Thus the evolutionary events of the latest Precambrian and Early Paleozoic produced a remarkable variety of basic animal groups, some of which are alive today.

In summary, what has unfolded with the many discoveries of the past two or three decades is a picture of rapid, but not instantaneous, diversification of life during the latest Precambrian and Early Cambrian—an interval in the order of 100 million years. Though perhaps divisible into pulses when scrutinized in detail, this general event constituted the most spectacular adaptive radiation of all time—the initial expansion of multicellular animal life. It is true, as often stated, that our ancestors lived in the sea, and it was in this great adaptive radiation that they got their start.

The new picture of rapid adaptive radiation relieves us of the need to invent unrecorded ancient worlds of life far back in the Precambrian, as Darwin and many later workers felt obliged to do. Not only is the need eliminated, but the recent focus on Precambrian and earliest Cambrian rocks, which has turned up the fossils just described, has revealed no unquestioned animals of much greater age. In other words, Precambrian life has finally come to light, but only in strata lying just below the Cambrian System.

It was a gradualistic view of evolution that led Darwin and others on their fruitless search far back into the Precambrian. Believing only in slow, persistent evolution, they had no alternative but to postulate a long, undocumented history of early animal life. Once again, the only refuge lay in condemnation of the fossil record. The record has now answered this challenge with solid evidence. The rapid adaptive radiation that is apparent today confronts gradualism with a seemingly insoluble problem. We now know that for many groups of marine invertebrates an average species lasts for five or ten million years without evolving enough to be given a new name. How, then, are we to explain the origin of advanced groups, like arthropods and mollusks, from primitive ancestors in a few tens of millions of years? Our only reasonable recourse is to abandon gradualism in favor of punctuational evolution, which can account for the rapid changes for which we see evidence. These changes must have been brought about by strongly divergent steps that came in rapid succession.

A parallel history of debate has surrounded the sudden fossil appearance of flowering plants. These so-called "angiosperms" form a coherent evolutionary group that includes not only plants with showy flowers, but also grasses and hardwood trees. Today their several hundred thousand species dominate the land. Flowering plants leave a legible fossil record primarily through the preservation of their leaves and pollen in volcanic ash deposits and in the sediments of river floodplains and lakes. This record has long been known to appear rather abruptly within the Cretaceous System. This pattern of appearance, like that of Cambrian marine life, confounded Darwin, whose letter of July 22, 1879 to Hooker referred to it as "an abominable mystery." Again Darwin reached in desperation for an ancient mythical kingdom:

> I have fancied that perhaps there was during long ages a small isolated continent in the S. Hemisphere which served as the birthplace of higher plants—but this is wretchedly poor conjecture.[9]

Younger workers sustained this fanciful attempt to provide the flowering plants with an early history adequate to permit gradual evolution to account for their complexity and variety in the Cretaceous Period. This effort continued into the 1960s, motivated in part by the periodic mistaken assignment of pre-Cretaceous fossils to the flowering plant group. Such assignments have all been deemed erroneous, however, and what remains is the original observation that a great variety of forms appear at approximately the same time in the Cretaceous. As with the Cambrian problem, it is the "approximately" that has recently been brought into focus.

In soft Cretaceous sediments of the Atlantic Coastal Plain between Washington, D.C. and Baltimore, James A. Doyle and Leo J. Hickey have uncovered evidence of the initial pattern of flowering plant diversification. Just as for the Cambrian faunas, what has emerged is a brief but discernible interval of adaptive radiation. A compelling aspect of the story is that it is told by both leaves and pollen. As one moves up through the geological column, the first leaves to be found are simple types that have smooth margins. Strangely irregular veins course through these leaves. Higher up are found increasingly complex leaves, some of which have toothed margins, and others of which are compound (occur several to a stem); the veins of these more advanced leaves follow more regular geometrical patterns, like those we are accustomed to seeing in the modern world. The earliest kinds of pollen match the earliest leaves in degree of simplicity. They have a single germination furrow, while pollen grains at higher levels are both more complex and more varied, with most species having three germination furrows and also increased surface sculpture of the sort that makes pollen grains sticky and easily transported by insects, such as bees.

The fossil record of flowering plants may extend a bit farther back in time than indicated by the Atlantic Coastal Plain occurrences, but not much. The oldest pollen and leaves of the Coastal Plain are not only limited in variety, they are primitive in form. What the record reveals is that considerable diversification from these early forms took place during just ten million years or so. This is not long by geological standards, but neither is it instantaneous. Darwin's abominable mystery is soluble, but here, as for the adaptive radiation of Cambrian marine life, the brief time scale opposes gradualism. The most plausible mode of change is one involving the rapid formation of new species. Thus, the fossil record has solved two century-old mysteries by revealing itself more fully, and in so doing it has spoken in favor of the punctuational model.

A third example, and one that lends itself to more detailed analysis, is that of the remarkable adaptive radiation of the mammals during the Cenozoic Era. The Age of Dinosaurs gave way to the Age of Mammals with striking abruptness. We still lack an explanation for the sudden extinction of the dinosaurs at the close of the Cretaceous, the last period of the Mesozoic. Whatever befell these great animals, my colleague Robert T. Bakker has conclusively shown that they were advanced and often highly mobile creatures, not lumbering behemoths that somehow became outmoded. They were, in fact, wiped from the earth while still quite successful—so successful that they had for some 130 million years suppressed the extensive adaptive radiation of mammals. The first mammals appeared on earth during the Triassic (the first period of the Mesozoic Era), but no mammal of the Mesozoic was much larger than a house cat, and it is commonly suggested that most small mammals that did exist were nocturnal. The fact that mammals underwent an extremely rapid adaptive radiation immediately after the dinosaurs died out represents strong evidence that

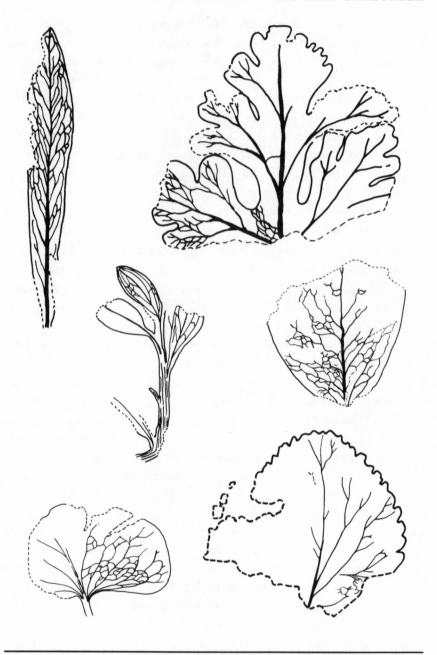

FIGURE 5.6
Fossil leaves of some of the oldest known flowering plants, showing primitive, poorly organized vein patterns.

the dinosaurs had dominated ecosystems to the point of blocking the diversification of mammals. It is uncertain whether it was through a competitive edge or through predatory behavior that dinosaurs maintained their dominance. Possibly both mechanisms were in operation.

When the mammals inherited the Earth, the result was spectacular. Their great adaptive radiation was recent enough that the fossil evidence for it is impressive. Within perhaps twelve million years, most of the living orders of mammals were in existence, all having descended from simple, diminutive animals that might be thought of as resembling small rodents, though not all possessed front teeth specialized for gnawing. Among the nearly twenty new orders were the one that contains large carnivorous animals, including modern lions, wolves, and bears; the one that comprises horses and rhinos; and the one that includes deer, pigs, antelopes, and sheep. Most of the orders evolved in even less than twelve million years. Perhaps the most spectacular origins were of the bats, which took to the air, and the whales, which invaded the sea.

Darwin was spared a confrontation with the extraordinarily rapid origins of modern groups of mammals. He knew that the history of mammals extended back to the early part of the Mesozoic, but the record was not well enough studied in his day for him to recognize that the adaptive radiation of modern mammals did not commence until the start of the Cenozoic. Today, our more detailed knowledge of fossil mammals lays another knotty problem at the feet of gradualism. Given a simple little rodentlike animal as a starting point, what does it mean to form a bat in less than ten million years, or a whale in little more time? We can approach this question by measuring how long species of mammals have persisted in geological time. The results are striking; we can now show that fossil mammal populations assigned to a particular Cenozoic lineage typically span the better part of a million years without displaying sufficient net change to be recognized as a new species.

The preceding observations permit us to engage in another thought experiment. Let us suppose that we wish, hypothetically, to form a bat or a whale without invoking change by rapid branching. In other words, we want to see what happens when we restrict evolution to the process of gradual transformation of established species. If an average chronospecies lasts nearly a million years, or even longer, and we have at our disposal only ten million years, then we have only ten or fifteen chronospecies to align, end-to-end, to form a continuous lineage connecting our primitive little mammal with a bat or a whale. This is clearly preposterous. Chronospecies, by definition, grade into each other, and each one encompasses very little change. A chain of ten or fifteen of these might move us from one small rodentlike form to a slightly different one, perhaps representing a new genus, but not to a bat or a whale!

What the gradualist must then postulate is an extraordinary acceleration of evolution within established species. In other words, he must claim that, in the lineage leading to the first bat or whale, chronospecies were actually of very short duration. This situation brings us to the essence of the gradualistic dilemma—a dilemma that holds for the adaptive radiations of Cambrian marine life and Cretaceous flowering plants as well. The first problem is that we have absolutely no fossil evidence for rapid transformation of chronospecies. On the contrary, early Cenozoic species of mammals appear to have had long durations, resembling those of younger species. The second problem relates not to fossil evidence, but to causal explanation. Why should well-established species suddenly undergo very rapid transformation? We know that after the demise of the dinosaurs the world was available for occupancy by mammals. Nonetheless, why should mere ecological opportunity cause any well-established species to abandon its way of life for an entirely new one? We might expect a broadening of the original way of life—of the niche, in the parlance of ecology—but not a desertion of what worked well before. Expanded ecological opportunity would be expected to permit great diversification, but no single species ever becomes very highly diversified. Rather, diversification proceeds by the sprouting off of new species from already established species—by adaptive radiation—and this, of course, brings us to the punctuational scheme of evolution.

Ecological opportunity sometimes exists on a grand scale. This happens when a part of the world is largely vacant, as the oceans were before multicellular animals first evolved in the Precambrian and as the continents were when dinosaurs vacated the Earth. Altogether new opportunities emerge in quite a different way when a new kind of biological feature (an adaptive innovation) evolves. The rapid adaptive radiation of the flowering plants is usually attributed to the advent of pollination by insects and birds. This system of reproduction is virtually unique to the flowering plants. Other land plants rely on less specific agents, like wind and water, to carry their pollen or sperm.

What is important to our assessment of gradualism is that adaptive radiations, which represent the exploitation of new opportunities, are the sites of most large-scale evolution. As described previously, nearly all of the major groups (orders) of the class Mammalia evolved during the first twelve or fifteen million years of the Cenozoic Era. During this very brief interval of geological time, the era became the Age of Mammals. The remaining fifty million years have seen subordinate groups (families, genera, and species) arise within these orders, but the basic parceling out of fundamental ways of life occurred early. Even the number of families reached its present level of slightly more than 100 about twenty-five million years after the start of the Cenozoic, although some families have arisen during the remaining forty million years of the era and others have become extinct. Certainly the most dramatic changes in mammalian evolution have been associated with adaptive radiation—with the initial radiation at the start of the era and lesser radiations within newer families. How long were the durations of established species while the initial mammalian radiation was in progress?

Superb fossil data have recently been gathered from deposits of early Cenozoic Age in the Big Horn Basin of Wyoming. These deposits represent the first part of the Eocene Epoch, a critical interval when many types of modern mammals came into being. The Bighorn Basin, in the shadow of the Rocky Mountains, received large volumes of sediment from the Rockies when they were being uplifted, early in the Age of Mammals. In its remarkable degree of completeness, the fossil record here for the Early Eocene is unmatched by contemporary deposits exposed elsewhere in the world. The deposits of the Big Horn Basin provide a nearly continuous local depositional record for this interval, which lasted some five million years. It used to be assumed that certain populations of the basin could be linked together in such a way as to illustrate continuous evolution. Careful collecting has now shown otherwise. Species that were once thought to have turned into others have been found to overlap in time with these alleged descendants. In fact, the fossil record does not convincingly document a single transition from one species to another. Furthermore, species lasted for astoundingly long periods of time. David M. Schankler has recently gathered data for about eighty mammal species that are known from more than two stratigraphic levels in the Big Horn Basin. Very few of these species existed for less than half a million years, and their average duration was greater than a million years.

The fossil species of Eocene Age are distinguished from one another primarily on the basis of their teeth—teeth being durable and especially well preserved in a fossil state. Most living species, and all genera, of mammals have distinctive patterns of dentition, which means that the interpretation of the fossils is generally reliable. The owners of a particular kind of teeth could not have undergone major evolutionary change in other ways while their dentition remained virtually unchanged.

It is ironic that among the sluggishly changing species of the Big Horn Basin were members of the "dawn horse" genus *Hyracotherium,* (formerly called *Eo-*

FIGURE 5.7
Early Eocene deposits forming badlands in the Big Horn Basin of Wyoming. These well-exposed sediments have yielded a remarkably complete record of mammalian species that lived here at a time when the climate was warmer and more humid than it is today.

hippus), the animal generally believed to be the distant ancestor of the modern horse. The fossil species of *Hyracotherium* show little evidence of evolutionary modification. One species lasted for at least three million years, and another for perhaps five million! For many years, while gradualistic thinking dominated evolutionary science, it was widely assumed that *Hyracotherium* had slowly but persistently turned into a more fully equine animal.

The new evidence for the stability of early Cenozoic species forces us to focus upon change by speciation involving small populations. Quantum speciation becomes our logical solution to the problem of the great mammalian radiation—a problem epitomized by the origin of bats and whales from small terrestrial mammals during twelve million years or less.

By studying newer groups of mammals, we can infer that the gradual modification of existing species cannot even account for the origins of most new genera. In particular, we can investigate rates of evolution during the Pleistocene Epoch—the famous Ice Age that extends backward nearly two million years from the present. It is generally agreed that we are still living within the Ice Age, having been granted a brief respite while the glaciers are in a state of recession. Continental glaciers have waxed and waned throughout the Pleistocene, and although exact predictions are at present impossible, we can look for the return of the ice sheets to the northern portions of the United States in several tens of thousands of years. The fossil record of mammals in sediments of the Ice Age is of superb quality. Teeth and bones are easily recovered from the readily accessible sand and mud laid down on Pleistocene floodplains, lake bottoms, and cave floors. The Ice Age faunas of thousands of collecting sites have been studied, and the result is a remarkably good picture of prehistoric life. Something like 85 percent of the species of mammals now living in Europe have been found fossilized in Pleistocene sediments. According to the

---

FIGURE 5.8
Skeleton of the "dawn horse," *Hyracotherium*. This animal, formerly known as *Eohippus,* was only about the size of a fox terrier. It had four toes on each front foot and three on each hind foot.

Finnish paleontologist Bjorn Kurtén, who has undertaken a comprehensive review of European Pleistocene mammal species, the few living species lacking Pleistocene records are species that one might expect not to find as fossils because they are either small, fragile forms or recent immigrants to the region. Of course a number of extinct species of mammals are found with the living species in Pleistocene deposits. In Europe, as well as elsewhere in the world, near the end of the most recent glacial episode a number of the largest land mammals, such as the woolly mammoth and woolly rhino, died out. This pulse of extinction is most commonly ascribed to the activities of human hunters. Let us focus, however, on the many more species that have survived to the present. These afford a special opportunity for examining rates of evolution within established species. In fact, the history of these living species shows that, since becoming fully established, they have evolved very slowly.

I have analyzed information on the apparent ancestries of the living mammal species of Europe. It turns out that, as we trace the populations of extant species backward through time, we find that more than half of the apparent lineages display such slow change that their million-year-old populations are grouped in the same species as their living populations. Even for the others, we have no certain evidence of gradual evolution. For the last third of a million years or so, a large majority of living species have left a fossil record, and here we find no species evolving sufficiently to be known by a new species name. All of this tells us that fully established mammal species—the kind that we see all around us—are evolving at an extremely sluggish pace. In this light, it is revealing to contemplate the origin of new genera.

Within the Mammalia, the genus is a rather discrete unit. Most genera seem to represent natural clusters of species, or discrete branches of phylogeny. How, we may ask, does a genus usually come into being? What is the mode of formation of its first species—the one from which other species of the genus are descended? The Pleistocene record of Europe exhibits several distinctive new genera. Among them are the following, most of which are represented today by a single species: *Thalarctos* (polar bear), *Rangifer* (reindeer-caribou), *Ovibos* (musk ox), *Capreolus* (roe deer) and *Microtus* (common vole). Some of these, like the polar bear, reindeer-caribou, and musk ox, are animals adapted to cold climates; their origins seem obviously related to the Ice Age climate.

Fossil data suggest that something like fifteen new genera evolved in Europe during the Pleistocene, or during less than two million years. Throughout this interval, at any one time, the total number of species (evolving lineages) numbered about 150. There were a number of extinctions, so that perhaps something like a hundred continuous pathways of gradual mammalian evolution traversed the Pleistocene in Europe. What happens if we attempt, hypothetically, to form each new genus by gradual modification along one of the recognized evolutionary pathways? What happens is that we are stymied! Genera differ enough from each other that a lineage that turned into a new genus would inevitably exhibit enough evolution along the way to be divided by a knowledgeable worker into several chronospecies. The problem that arises is that species of the Pleistocene encompass so much time that we would be unlikely to encounter any lineage that an expert would divide into more than three or four successive chronospecies within the 1.8 million-year Pleistocene Epoch (certainly none is known). This does not get us very far. Even the amount of evolution marked off by as many as five or six successive chronospecies might not give us a new genus.

In the midst of all this sluggish change within established Pleistocene lineages there cropped up strikingly distinctive new genera. The polar bear, for example, has entered a new kind of ecological niche for the bear family, living a semiaquatic existence and feeding almost exclusively on seals. Its teeth are modified for pure carnivory. It is not to be taken lightly that, despite living in an aquatic habitat where preservation of bones is favored, the polar bear has almost no fossil record. Possible remains are few and are known only from sediments dated at forty thousand years or less. To explain the appearance of the polar bear and most other new genera of the Pleistocene, we seem to have no choice but to invoke the rapid divergence of populations too small to leave legible fossil records. Some genera may have formed by two or more rapid speciation events of this kind. A single step of rapid branching is, however, probable for some, such as the polar bear, which clearly formed from the brown bear (the species *Ursus arctos,* which contains several varieties, including the European brown bear, the grizzly, and the Kodiak).

The evolution of the elephants presents us with a well-documented illustration of the pattern I have described. In recently reconstructing the family tree of this group, Vincent Maglio concluded that the fossil record of the modern subfamily Elephantinae is now very well known. Elephants are so large that their bones and teeth are readily preserved and discovered. Recall that because living elephant species also are not easily overlooked, Cuvier turned to the elephants in order to demonstrate the reality of extinction. (An indication that indeed the elephant record is now well known is the fact that in recent years very few new kinds of fossil elephants have turned up.) During the past five million years, three genera of elephants have populated the earth. All are familiar. One includes the living African elephant, one the living Indian elephant, and one the extinct mammoths. The history of each genus extends back more than four million years. During most of this time the genera comprised several species. Our two lone survivors, the African and Indian elephants, are the vestiges of a once mighty group. What stands out in the phylogeny of the elephants is a pattern strongly indicative of punctuational evolution. All three advanced genera descended from the ancestral genus *Primelephas,* and all three appear abruptly and almost simultaneously in the fossil record. The subsequent history of each spans about four and one-half million years, and while during this long interval certain paleontologists have asserted that some gradual evolution took place, no lineage of any elephant genus changed enough to represent a new genus. The genera, once formed, retained their basic body plans through something like half a million generations.

The Ice Age, with its pronounced environmental vicissitudes, hardly represents a time when one would expect evolution within established species to have slowed greatly from its normal pace. For this reason, we seem fully justified in extending our punctuational findings to the evolutionary history of mammals in general. Certainly, we must ask, if gradual transformation within established lineages cannot account for the origins of most genera, how can it be expected to account for the origins of most families or orders? In fact, the data for the Pleistocene mammal species are compatible with those for the older mammal species of the Bighorn Basin—species that underwent almost no perceptible evolution early in the Cenozoic Era, while strikingly new types of mammals were appearing.

While groups other than the mammals are not amenable to such detailed analysis, they provide data that point in the same punctuational direction. The Foraminifera represent an example. These are tiny, single-celled, amoebalike organisms with shells. "Forams" of the sort that live in sediments on the sea floor are represented in many handfuls of beach sand by dozens of dead shells. Such shells are not only abundant, they are also readily fossilized. They are used extensively for dating sedimentary deposits in the search for petroleum because they are accessible within the small chips and cores recovered from drilling operations. An average species of sea-floor dwelling foram alive at any time during the Cenozoic survived for more than twenty million years (at least one-third of the entire era!) with little change. This seems a remarkably long time when we consider that more than twenty distinctive families of forams appeared during the first thirty million years of the era—in other words, during an interval not much longer than the average duration of one species! Apparently, to explain the origin of a family, we must look to some mechanism other than the transformation of established species.

The bivalve mollusks comprise clams, mussels, oysters, scallops, and their relatives. If we scrutinize the Cenozoic fossil record of marine bivalves, we find that about half of all species recognized in marine fossil faunas seven million years old are also recognized in the modern world. Many of those not alive today belong to lineages that have been terminated. Had no lineages died out, even more than half of the chronospecies seven million years old would be recognized in modern seas. In other words, most bivalve chronospecies last much longer than seven million years!

Example after example strengthens the case that rapid evolution is restricted to small populations. For large populations of the sort that constitute the species living all around us today, sluggish change is the norm—much more sluggish change than all but a few modern biologists have envisioned.

Since undertaking tests of gradualism, I have been surprised to find that many of the gnawing questions that have motivated me also motivated others long ago, but to no avail. Since the time of Darwin, paleontologists have found themselves confronted with evidence that conflicts with gradualism, yet the message of the fossil record has been ignored. This strange circumstance constitutes a remarkable chapter in the history of science, and one that gives students of the fossil record cause for concern.

Consider the fate of the German paleontologist Heinrich-Georg Bronn, whose career overlapped with Darwin's. In 1857, Bronn received the Grand Prix for Physical Sciences in a contest of the Academy of Sciences of Paris for essays that considered the temporal occurrence of fossil forms and their relationship to living species. The motto of Bronn's contribution was "To be taught by nature." What nature taught him was that species survive for a long time. Discerning no evidence of any change whatever in fossil species, he became a believer in the immutability of species. This view was extreme, but Bronn was nonetheless an expert on Cenozoic marine fossils, and two years later, when the *Origin* appeared, his evidence represented a valid challenge to Darwin's gradualism. Lyell, who long shared Bronn's opinions, was also an expert on Cenozoic marine life and, like Bronn, he was struck by the persistence of many species through sedimentary units that seemed to represent considerable spans of time. To these men, the Cenozoic record spoke convincingly, as it does to me. Bronn died within three years of the *Origin*'s publication, so that his opposition was short-lived. His eminence faded into obscurity, just as the reputation of paleontology in general declined when it became evident that the fossil record would not, in detail, support Darwin's thesis. Few living paleontologists recognize the name Heinrich-Georg Bronn.

Other paleontologists challenged Darwin in the immediate aftermath of his provocative publication. In 1860, John Phillips, president of the Geological Society of London, presented a lecture "Life on the Earth: Its Origin and Succession," in which he offered expert testimony that the fossil record was indeed of high enough quality to test Darwin's notion of widespread gradual change. He then claimed that the facts denied Darwin's conception of evolution. In particular, he cited the first example I have described: the sudden appearance of advanced, varied forms of life at the beginning of the Paleozoic.

I have been interested to discover that one early student of fossils focused upon the elephants, which as especially well-preserved mammals have formed an important part of my case. This early worker was Hugh Falconer, who in 1863 published an article on the elephant fossil record and who also received letters from Darwin from which I quoted in the previous chapter. Falconer observed that elephants of the Ice Age exhibit little temporal change in their dental or locomotory features. He was particularly puzzled by the stability of the mammoth:

> The whole range of the Mammalia, fossil and Recent, cannot furnish a species which has had a wider geographical distribution, and at the same time passed through a longer term of time, and through more extreme changes of climatal conditions, than the Mammoth. If species are so unstable, and so susceptible of mutation through such influences, why does that extinct form stand out so signally as a monument of stability?[10]

As I noted in the previous chapter, Falconer chose not to defy Darwin on this issue. Instead he suggested that perhaps natural selection operated most effectively when species first form. This prescience had greater implications than he or Darwin knew. In editions of the *Origin* published after 1863, Darwin paid lip service to Falconer's hypothesis, with proper attribution, but was never truly disengaged from his gradualistic position. The rest of the *Origin* remained heavily gradualistic. My real point, however, is that elephants, which loom large in my argument, offered up their conspicuous evidence at a remarkably early date.

T. H. Huxley, though Darwin's most vocal champion, remained throughout life a skeptic of the efficacy of natural selection. Huxley was both a zoologist and a paleozoologist. Major addresses that he presented in 1862, 1870, and 1880 reveal the history of his discomfort with gradualism—a discomfort engendered largely by his considerable knowledge of fossils. Huxley's Anniversary

The Fossil Record: Our Window on the Past

Address to the Geological Society of London in 1862 concluded with a summary statement on fossil evidence with respect to "doctrines of progressive modification."[11] "It negatives those doctrines," he claimed, "for it either shows us no evidence of any such modification, or demonstrates it to have been very slight." Like others, Huxley was particularly moved by the apparently sudden appearance of complex life at the start of the Paleozoic. In his address to the same body in 1870, Huxley admitted evidence of step-by-step transitions only for Cenozoic mammals, and these transitions were no more than crude series of genera—primarily the series suggested by the Frenchman Albert Gaudry. They were not approximately continuous lineages. Huxley also amplified his perplexity over a problem that he had mentioned in his address of 1862—a problem that foreshadowed what I have called "the test of living fossils." He was not concerned in particular with the absence of major transitions when there was little speciation; he was simply confounded by the existence of "persistent types," or body plans which changed little through the ages:

> The significance of persistent types, and of the small amount of change which has taken place even in those forms which can be shown to have been modified, becomes greater and greater in my eyes, the longer I occupy myself with the biology of the past.

Huxley was also disquieted by the rapid appearance in the early Cenozoic of many orders of mammals. He was particularly struck by the early origins of the bats and the whales, the very groups that first arrested my attention when contemplating the punctuational model, though I had never seen Huxley's century-old address. I am not boasting here, because in truth there is nothing subtle about these observations; the examples jump out at us as unmet challenges to gradualism.

Huxley, lacking the broader knowledge at our disposal today, sought refuge where Darwin himself had hidden. Huxley postulated fantasy lands in which long, unrecorded evolutionary histories might have unfolded. Rather than relying solely on evolution by rapid steps (an idea that he had earlier proposed informally to Darwin), he now sought to preserve gradualism by granting it more time. He stated that "if there is any truth in the doctrine of evolution, every class must be vastly older than the first record of its appearance upon the surface of the globe." He then envisioned a lost continent of Late Paleozoic and Mesozoic Age, upon which the many advanced mammal groups of the Early Cenozoic might have evolved in a gradual fashion before making their way to Eurasia and America. We can now rule out this fanciful scheme. Large conti-

nents cannot have foundered and disappeared altogether during the Phanerozoic (continents are buoyant with respect to the dense rock of the Earth's
mantle, on which they rest). Furthermore, we find no mammalian fossils of
advanced Cenozoic grade in the Mesozoic rocks of any existing continent.
Again, it is not Huxley's hypothetical solution that is important, but the problem itself. How did advanced mammals arrive on the scene so quickly in the
Cenozoic?

In 1880, Huxley presented a paper that was more optimistic in both content
and title: "The Coming of Age of the Origin of Species."[12] By this time,
Huxley had become a believer in evolution. What had moved him?

> Simply this, that, if the doctrine of evolution had not existed, paleontologists
> must have invented it, so irresistibly is it forced upon the mind by the study of
> the remains of the Tertiary [Cenozoic] mammalia which have been brought to
> light since 1859.

Huxley had been impressed by a variety of mammalian evidence, but in
particular, when he visited the United States in 1876 to speak at the founding
ceremony for the Johns Hopkins University, he took the opportunity to visit
O. C. Marsh at Yale. Marsh's collections convinced the previously skeptical
Huxley that the horse family had evolved in North America. What had previously fooled Huxley was an occurrence that I mentioned earlier. Horses had
spread to the Old World across the Bering Land Bridge, which at certain times
connected Alaska and Asia, and then died out in the New World. Before their
New World fossil record was known, they seemed to be native to the Old
World, and it appeared that they had been brought for the first time to the
New World by the Spaniards.

Huxley's belated enthusiasm for evolution is not surprising. The mammalian
fossil record was, and is, a compelling source of evidence for the reality of
evolution. What is unfortunate is that the momentum of his conversion caused
him to gloss over the sudden fossil appearances and "persistent types" that had
troubled him in the past and that might have led him to pursue his earlier
objection to Darwin's claim that life does not change by jumps. Huxley out-
lived Darwin by thirteen years, dying in 1895. Toward the end, he gave little
voice to his negative attitudes toward gradual change and natural selection.
Evolution, in general, was entering troubled times and presumably, as a believer, Huxley was not inclined to aid those who were disposed to throw the baby
of evolution out with the bathwater of gradualistic natural selection.

The softening of Huxley's voice on the problems raised by fossils epitomizes what happened to paleontology late in the century. The fossil record offered convincing evidence for evolution, but not for the prevalence of *gradual* evolution. Unfortunately, circular reasoning soon crept into the evaluation of the record. Darwin had made elaborate claims that fossil data were too sparse ever to support his gradualistic scheme, yet his condemnation of the record was not based on objective observation. About this, he was quite open in the *Origin* (p. 302):

> But I do not pretend that I should ever have suspected how poor a record of the mutations of life, the best preserved geological section presented, had not the difficulty of our not discovering innumerable transitional links between the species which appeared at the commencement and close of each formation, pressed so hardly on my theory.

In other words, Darwin had deduced the incomplete nature of the fossil record from his theory. This approach was uncharacteristic. Darwin saw himself as following the inductionist tradition of Francis Bacon, gathering large numbers of observations until they converged on some idea. What Darwin actually meant was that he was an empiricist, an observer who gathered facts in order to generate theory. He condemned "deduction," by which he meant speculation. Undoubtedly his position represented a reaction against the habit that his predecessors had of deducing all sorts of consequences from unfounded postulates, such as the existence of a *Scala Naturae*. Certainly, Darwin often employed deduction legitimately: normally he deduced consequences from solid observations or from hypotheses that he wished to test.

In deducing the poor quality of the fossil record rather than studying fossil data in detail, however, Darwin violated his commitment to empiricism. The first edition of the *Origin* contained very little paleontological information (few relevant data had been published). Later editions incorporated several examples of fossil data that had come to light, suggesting some form of evolutionary descent. What, in particular, were missing throughout Darwin's lifetime were well-documented and well-publicized examples of fossil species having great geological longevity. Although a few workers, like Bronn and Falconer, had adequate experience to argue for the presence of such fossil species, living before the advent of radiometric dating, they had no way to support their position with quantitative data. Lyell's powerful support of Darwin's deductive denigration of the fossil record was, as I have noted, based on personal bias: for years he had denied that fossils even attested to a general progression of basic life forms. As M. J. S. Rudwick[13] has observed, it was in large part because fossils were seen as offering little information about the nature of the evolutionary process, that, after publication of the *Origin,* paleontology fell from scientific prominence for many years. Darwin himself concluded in the *Origin* (p. 487):

> The noble science of Geology loses glory from the extreme imperfection of the record. The crust of the earth with its embedded remains must not be looked at as a well-filled museum, but as a poor collection made at hazard and at rare intervals.

It was soon forgotten that Darwin's judgment of the fossil record was based on deduction rather than fact. It came to be assumed that there was solid evidence of gaps on a critical scale. This amounted to a kind of historical circularity of reasoning. The absence of an absolute scale of geological time in the nineteenth century can not be overemphasized here. Only because we can approximately date geological intervals can we now show that species range over long spans of time with little evolutionary change. We can also show that, while there are gaps in the record, in many places the gaps are on a scale that does not interfere with our ability to trace out many lineages by way of sequential, though not always contiguous, faunas, and to observe degrees of change.

As I have previously noted, some paleontologists continued to raise objections to gradualism until well into the twentieth century. The problem was that, as before, they tended to reject natural selection along with gradualism. The two had been linked unnecessarily by Darwin, but the bond was not easily broken. There was little consideration of the possibility that selection might prevail, but as Hugh Falconer bootlessly suggested to Darwin, its effects might be concentrated in small populations rapidly forming new species. Thus, when geneticists finally reasserted gradualism in the 1930s and soon thereafter came to dominate evolutionary biology, paleontology remained in a peripheral role. It had become one of the evolutionary fields that before the turn of the century were scorned for being at once speculative and boring. Lacking a gradualistic or selectionist pedigree for entry into the Modern Synthesis, paleontology could not easily gain in status during the 1930s.

In truth, it seems fair to say that, to the degree that paleontology joined the Modern Synthesis at all, it was as a stepchild. The Foreword of *Genetics, Paleontology, and Evolution,* a book edited by Glenn L. Jepsen,[14] Ernst Mayr, and George Gaylord Simpson, reveals some of the relevant history. The book itself was a document prepared for the National Research Council's Committee on Common Problems of Genetics, Paleontology, and Systematics. It recorded the events of a symposium held at Princeton University in 1947, but the origins of the symposium itself trace back to 1941, when at the annual meeting of the Geological Society of America, Walter Bucher suggested that the divergent evolutionary fields of paleontology and genetics should be united. Beginning in 1942, meetings were held to explore the matter, and the formal committee was appointed in 1943. The Society for the Study of Evolution was formed in 1946, and its journal, *Evolution,* was founded in order to display the products of the union. It was in this way that paleontology was brought under the umbrella of the Modern Synthesis: by the action of a committee. This was a shotgun wedding, not a common-law marriage. Immediately after its inception in 1947, the journal, *Evolution,* embraced evolutionary subjects representing many fields, but soon its emphasis shifted away fron paleontology and toward genetics. The old polarity surfaced again.

Tempo and Mode in Evolution[15] was published in 1944. It was an important book, written by the paleontologist George Gaylord Simpson. *Tempo and Mode* was begun in the spring of 1938 and, although published in 1944, it was out of Simpson's hands well before this date. In timing and in spirit, its preparation represented the era that preceded the main activities of the National Research Council Committee, which Simpson chaired. *Tempo and Mode* was an imaginative and daring book, an important aspect of which was its use of fossil data to study rates of evolution. In the present context, what was most important was that Simpson's book was not entirely gradualistic. Simpson proposed the concept of quantum evolution, which he defined as very rapid evolution within a small population, leading to a dramatically new kind of animal or plant—primarily a new family or order. Simpson postulated that this sort of evolution is initiated by the random fixation of initially inadaptive mutations, followed by selection that establishes a new adpative relationship to the habitat. Simpson's quantum evolution differed from the kind of rapid change that I will discuss in the next chapter, in that quantum evolution was not linked specifically to branching (speciation). Simpson asserted that a distinctive family or order of animals or plants might arise by way of an inadaptive phase, and this ran counter to important tenets of the emerging field of modern genetics. Genetics emphasized gradualism and the tight control of adaptation by selection.

In reviews of *Tempo and Mode,* quantum evolution was criticized as heretical, yet the concept had a familiar source. Simpson was one more paleontologist struck by the suddenness of change in the fossil record. Once again, however, the biologists refused to relinquish their gradualistic beliefs. After the committee, formed to unite paleontology and genetics, had finished its work, Simpson published a refurbished version of his book, entitled *The Major Features of Evolution*.[16] Here, not surprisingly, large sections on genetics were added. Only the final three and one-half pages were devoted to the concept of quantum evolution, compared to more than twelve pages in *Tempo and Mode*. The thesis of the second book was predominantly gradualistic.

In 1950, the prominent German paleontologist Otto Schindewolf[17,18] published two books that were shockingly anti-Darwinian. Schindewolf believed that the fossil record demanded a mechanism of sudden change. He adopted the old idea of macromutation, envisioning such things as the first bird hatching, fully formed, from a reptile egg. Schindewolf was widely derided in English-speaking countries. In his second book, Simpson was forced to refute many of Schindewolf's ideas, but he was careful to state that Schindewolf's work was grounded in an excellent knowledge of the fossil record. This point bears emphasis. While Schindewolf resorted to unnecessarily radical mechanisms, he, like less extreme opponents of gradualism, was attempting to contend with hard facts.

Although in 1954 Ernst Mayr[19] published his paper suggesting that "evolutionary novelties" often evolve rapidly in localized speciation events, the momentum of genetics continued to prevail. Punctuationalism went underground for nearly two decades. It was not until 1971 and 1972, when Niles Eldredge and Stephen Jay Gould[20,21] reiterated Mayr's arguments on speciation, that biologists finally began to take notice. By this time, the Modern Synthesis had weakened slightly with age. The centennial celebration of the *Origin of Species* was more than a decade behind us, and geneticists were reshaping their science in terms of regulatory genes, which I will describe in the chapter that follows. In 1970 Ernst Mayr wrote, "The day will come when much of population genetics will have to be rewritten in terms of the interaction between regulatory and structural genes."[22] The truth is that regulatory genetics played into the hands of punctuationalism. The time was ripe for change.

The history of science endows paleontology with credibility as an arbiter in the controversy over gradualism. Certainly genetics is imbued with mathematical rigor, and quantitative disciplines customarily carry prestige in the pecking order of science. Because many fundamental assumptions must be employed in making nature mathematically tractable, however, heavily quantitative analyses have often led earth science astray. Two examples are the mathematical arguments that needlessly troubled Darwin: the claim that blending inheritance undermined the potential efficacy of natural selection and the calculation that the earth was very young because it had to be cooling rapidly. Another case in point is the calculation by geophysicists early in this century that continents could not drift over the Earth's surface because they were supposed to be locked in place by rigid crustal rock beneath the oceans; we now recognize that the oceanic crust, though rigid, slides right along with the continents, but until this was shown, an enormous array of geological evidence favoring continental movement was held at bay.

The lessons of history provide us with a case for favoring strong circumstantial evidence of the geological record. If rocks give evidence of having formed slowly, then the Earth is very old. If coastlines and other geological features show close similarity across ocean basins, then distant shorelines were once in apposition: continents have rifted apart. Finally, if fossil species survive for long intervals with only minor modification, then the gradualistic model of evolution is denied; most change must occur in small populations.

Geology, particularly its paleontological aspects, had been the most popular of natural sciences in England during the two or three decades preceding Darwin's publication of the *Origin of Species*. This was true in an avocational sense, but in those days there were few professional scientists in any field, and for doctors, ministers, and gentleman farmers, amateurism quickly became hard science. After 1859, all this changed. Unfortunately, the crystallization of Darwinian gradualism and the accompanying misguided attack upon the quality of the fossil record sent paleontology into a major decline. In the words of Martin J. S. Rudwick, a paleontologist turned historian of science, "Only recently have there been hopeful signs that paleontology may be recovering, in its younger generation, the broad interests and outlook that it possessed so markedly earlier in its history." [23]

On the Rapid Origin of Species

E VEN if Darwin had possessed theoretical reasons for adopting a punctuational model of evolution, for him to have advanced such a scheme would have seemed as absurd as were many of the speculative pronouncements of his predecessors. As I noted earlier, he would have been claiming that evolution, a process not widely believed in, not only was real but was operated by a natural mechanism whose major effects are wrought exactly where we are least able to study them—in small, localized, transitory populations. Scientific ideas that are untestable when proposed may be correct but are seldom popular. How can they be?

Darwin's opposition to punctuational evolution was not overcome by facts because facts were not available. One deficiency, of course, was the generally poor state of knowledge of the fossil record. This problem diminished only with the slow accumulation of information. Even fifteen years ago, it would have been difficult to fashion published fossil data into a picture of chronospecies longevity. Also lacking was evidence that distinctive species can actually form rapidly from small populations. The information we now have on this topic forms the subject of this chapter.

We have seen how fossil data have come to demand rapidly divergent speciation. We will now see that this demand is not biologically unreasonable. Year

On the Rapid Origin of Species

by year, the geography and the genetics of life point more and more convincingly towards the occurrence of the extraordinary events that our model requires.

For an example of speciation to fit our bill, it must meet two criteria. First, it must have occurred rapidly—in less than a few thousand years or at most a few tens of thousands of years. Second, it must have achieved significant change—the degree of change that might represent the formation of a new genus or the achievement of a large step toward one. What we seek, then, are evolutionary Gardens of Eden. Unfortunately, the Garden of Eden of a particular species tends to disappear with time or to become unrecognizable as the species expands to new regions. What we must locate are species that have formed recently and by evolution within small populations that have not yet escaped their places of origin. Not long ago, such species were all but unknown, but during the past few years, a number have come to light.

Lake Nabugabo is a small body of water in Uganda, at the margin of Lake Victoria. Geological evidence shows that Nabugabo, which measures only about three miles by five, formed by the growth of a sand spit across a small embayment at the margin of Lake Victoria. Radiocarbon dating of plant materials incorporated into the spit reveals that closure occurred quite recently, approximately four thousand years ago, at a time when, to the north, civilization had already come to the Middle East. The remarkable fact is that Lake Nabugabo harbors five species of cichlid fishes that are unknown anywhere else, including Lake Victoria, but each of which resembles a species of the ancestral lake. The Nabugabo species differ from these obvious parent species of the larger lake in the coloration of male animals and in other minor traits. Male color patterns represent an important feature in the identity of species, serving as a badge of recognition for breeding. Here we have an obvious example of rapid speciation. Small populations of the parent species of Lake Victoria were apparently isolated when the spit sealed off Nabugabo as a separate lake, and they have since become new species. The four-thousand-year age of Nabugabo establishes a maximum interval for the origin of its unique species; most of their distinctive features may actually have formed much more quickly. Still, even four thousand years is a brief instant in geological time, and also a brief instant with respect to the evolution of freshwater fishes, an average species of which has lasted nearly a thousand times this long—something like three million years.

In Hawaii, there are several species of moths of the genus *Hedylepta* that feed exclusively on banana plants. All other Hawaiian species of this genus feed on grasses, sedges, lilies, palms, or legumes. The notable fact here: bananas

FIGURE 6.1

Two Hawaiian species of banana-feeding moths. These species have originated during the past one thousand years or so, since the banana was introduced to Hawaii. Above: *Hedylepta maia.* Below: *Hedylepta meyricki.*

were introduced to the Hawaiian Islands by Polynesians only about one thousand years ago! *Hedylepta* has apparently experienced multiple speciation events during this brief time to develop species that feed only on the newly arrived banana. Each banana-feeding species of moth is confined to mountain forests on only one or two islands. It is understandable that the species in question are restricted in distribution. They have not yet spread far from their cradles of origin.

The fishes and moths just described have had quite recent origins, but none-

On the Rapid Origin of Species

theless origins that have not entailed dramatic divergence. In no case has a new genus formed or a large step been taken in this direction. The pupfishes of Death Valley, along the California-Nevada border, do satisfy this second condition. One of these, *Cyprinodon milleri,* was not discovered until 1967. It is a minnow-sized animal confined to waters associated with Cottonball Marsh, a small complex of aquatic environments where temperatures range from tepid in summer to near freezing in winter and where the dissolved salt content of the waters also varies sharply. Geological evidence reveals that Cottonball Marsh came into being no more than a few thousand years ago, yet *Cyprinodon milleri* is quite distinct from other species of its genus. Among its unusual traits are distinctive teeth and an almost complete lack of pelvic fins. This well-defined species apparently evolved rapidly from *Cyprinodon salinus,* which inhabits nearby bodies of water and seems to be its closest relative.

Cyprinodon milleri is not the only distinctive and youthful species of pupfishes in the vicinity of Death Valley. Others occupy localized bodies of water in the desert—oases for aquatic life that can be no older than ten to thirty thousand years. The Death Valley region has not always been hot and dry. During intervals of the recent Ice Age when glaciers advanced over the northern part of the North American continent, climates to the south were in many areas much wetter than today. Death Valley is such an area. While, between ten and thirty thousand years ago, the last continental glaciers receded northward, Death Valley became a desert, as presumably it had been during earlier interglacial stages. A number of small pupfish populations remained in scattered springs and streams. Several of these diverged rapidly to become distinctive new species. One of these is the devil's pupfish, so named because it lives in water at a temperature of 92°F. Its home is a tiny thermal spring, where its entire population has probably never expanded beyond 300 individuals! Had this and other isolated species lived before the last glacial age, their populations would today enjoy more widespread distribution in western North America. Clearly, they evolved their distinctive features only after being isolated by the recent drying up of the Death Valley region. The devil's pupfish is very unusual. On the basis of form, it constitutes a new genus, although it has not yet been christened with the unique generic name it deserves. Like *Cyprinodon milleri,* the devil's pupfish has reduced pelvic fins or none at all, but it is a more distinctive animal, bearing no close resemblance to any other living species.

Today the devil's pupfish is gravely threatened. The waters feeding its spring, and those that supply neighboring springs that sequester other rare fishes, are now being diverted by humans for irrigation. Several fishes of the region have been driven to extinction, and others are listed as endangered species. The devil's pupfish is sustained precariously by water covering a single narrow rock ledge; here it lays its eggs and algal growths supply its food. If the local water table declines much further, the species will almost certainly disappear. Fortunately, in 1976, the United States Supreme Court ruled in this small animal's favor, enjoining landowners from siphoning off any more of its life-giving waters.

We must not be misled by those who have irreverently enshrined the snail darter as the symbol of inconsequential life. Many humble and inconspicuous species are of great value to science. The tiny devil's pupfish is one of a modest number of species in the world that may hold the key to rapidly divergent speciation. For its genes, its physiological mechanisms, its tissues, and its behavior to vanish from the Earth along with its habitat would be tragic. Such a loss would deprive us of the rare opportunity to examine the particular process by which a distinctive new species has come quickly into being. As human depredation of the global environment proceeds, it is not the visibly important, well-established species that we endanger, but the populations confined to small areas, and unfortunately it is only among these that we can hope to catch species in the act of forming.

Small bodies of water are also known to harbor flocks of unique fish genera that have come into being quite recently. One of the most spectacular of these is Barombi Mbo, a lake that occupies the crater of a volcano in Cameroon. Barombi Mbo is less than three miles in diameter, and the volcano in which it sits is only at most a few hundred thousand years old. Seventeen species of cichlid fishes inhabit the lake, and twelve of these are endemic (unknown from any other locality). Seven of the endemic species belong to four endemic genera. These genera are so unusual that it is not even clear exactly where their ancestry lies among the cichlid species of the surrounding area.

Just as Darwin and many architects of the Modern Synthesis would have been confounded by the fossil evidence for the relative stability of most well-established species, they would have been shocked at how rapidly small populations can diverge to form new genera. Whereas it is now arguable that only a small fraction of all genera have formed by the gradual transformation of full-fledged species, Darwin viewed genera as forming *only* in this manner. In Chapter Four of the *Origin,* he diagrammed the tree of life as having branches that diverge gradually. For him, the typical pattern of divergence entailed one population of a species becoming sufficiently altered in the course of many generations to be recognized as a new variety and, later, a new species; he believed that

On the Rapid Origin of Species

eventually sufficient change would accumulate in the new lineage thus formed that before becoming extinct it might well be recognized as a new genus, or even a new subfamily. In 1962, Theodosius Dobzhansky wrote, "The core of Darwin's argument was that species arise from races by a process of gradual divergence,"[1] and ten years later, Dobzhansky himself described the gradual formation of species as the "usual, and by now orthodox, view."[2] What was wrong here was not the notion that species can come into being slowly, but the notion that nearly *all* species form in this manner. Very little attention was paid to the kinds of rapid speciation events that now seem so important.

For convenience in describing rapidly divergent speciation, we now need a special name. The one that has taken hold, "quantum speciation," was supplied in 1963 by the botanist Verne Grant, an early advocate of the modern punctuational view. The adjective here was borrowed from George Gaylord Simpson's phrase "quantum evolution," which referred to any kind of rapid evolution in a small population, including the passage of a well-established species through a bottleneck from which it emerged as a different form. Our numbers now seem to show that this kind of rebirth of a species by "bottle-necking" is unlikely. For every species transformed in passing through a bottle-neck, there must be dozens that fail to emerge—that become extinct. We can calculate the number of extinctions of species that actually occur for particular groups of animals within fixed intervals of time. As it turns out, this number is too small to permit us to envision species to be reborn by "bottlenecking" except very rarely. As we shall see, most small populations available for rapid transformation are not the solitary remnants of once-thriving species, but discrete populations of still large and successful species; at any given time, large and successful species include many small populations, any one of which may blossom into a distinctive new species.

How is it that species form rapidly? Our evidence points to transitions within small, localized populations, but what mechanisms are involved? Indications are provided by examples of multiple speciation events that have occurred recently within known areas: adaptive radiations whose products are still alive and available for study more-or-less where they came into being.

The Galápagos finches are famous for their role in Darwin's thinking, but the divergence that they represent is by no means extraordinary, and Darwin actually did not study them with care. Far more dramatic was another adaptive radiation of birds, also issuing from finch ancestors, on the Hawaiian Islands. This episode gave rise to the so-called honeycreepers, which are found only in the Hawaiian archipelago and which display an extraordinary variety of beak

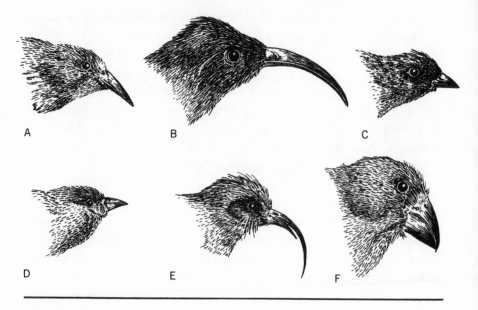

FIGURE 6.2

Variation of bill shape among Hawaiian honey creepers. A: *Himatione sanguinea* (nectar feeder). B: *Drepanis pacifica* (nectar feeder). C: *Ciridops anna* (fruit eater). D: *Loxops coccinea* (insect eater). E: *Hemignathus lucidus* (insect eater). F: *Psittirostra palmeri* (fruit and seed eater).

shapes. A number of honeycreepers have parrotlike beaks. Others have long, curved beaks, which are slender in some species and robust in others. Still others have small, nondescript beaks resembling those of most song birds. Some honeycreepers feed on the nectar of flowers, others on insects, and still others on seeds or fruit. Several dig insects from wood. The Hawaiian Islands, like the Galápagos, are geographically youthful volcanic constructions. As a group, the Galápagos are less than three million years old, but particular islands are not well dated. More precise radiometric dates are available for the individual Hawaiian Islands. Kauai, the westernmost large island and the oldest, is dated at about 5.6 million years. Hawaii, the easternmost island and the youngest, is only about 750 thousand years old. A number of honeycreeper species are endemic to the youthful island of Hawaii, or were before becoming extinct. This indicates that much of the adaptive radiation of honeycreepers has occurred very recently and, in fact, fossil evidence gathered by Storrs Olson and Helen James of the Smithsonian Institution suggests that much speciation has taken place during the latter part of the Ice Age. Regrettably, because of hu-

116

man destruction of native Hawaiian habitats, a number of honeycreeper species have died out, and others are on the brink of extinction.

The Hawaiian Islands also happen to have been the site of a prodigious, recent adaptive radiation of drosophilid fruit flies, the group made famous by thousands of laboratory experiments in genetics. The Hawaiian archipelago is populated by perhaps as many as 800 species of *Drosophila,* a number of which have evolved on the youthful island of Hawaii.

Many African lakes, such as Barombi Mbo, which was described earlier,

---

FIGURE 6.3
Three types of cichlid fishes from Lake Victoria, Africa: an insect eater (top), a fish eater (center), and a mollusk eater (bottom).

harbor diverse flocks of cichlid fishes. A particularly interesting example is Lake Victoria, the parent body of water of Lake Nabugabo. Lake Victoria, like the island of Hawaii, is only about 750 thousand years old, yet its fish fauna comprises approximately 170 cichlid species found nowhere else! For a fresh-water lake, Victoria is enormous, having an average diameter of more than 200 miles. The surrounding streams and rivers support only a few cichlid species. Interestingly, these are generally types that seem to represent the ancestral con-dition of the Victoria species flock, which in contrast to the sparse faunas of the streams and rivers includes a spectacular variety of adaptive types. Some species of the lake feed on insects, some eat other fishes, several devour only fish larvae and embryos, a few subsist on mollusks, and a number graze on plants. The diet of one species consists in large part of the scales of other fishes! The cichlids exhibit a wide range of mouth shapes and dental features that in most cases show a close association to feeding habits. For the time being, nearly all of the 170 species have been left within the ancestral genus *Haplochromis,* but, on the basis of diversity of form and behavior, the species should be divided among a number of genera.

As to the way in which the cichlid flock of Lake Victoria came into being, a telling fact is that there reside within the lake a few species that seem to represent the ancestral condition for the group. These are insect-eaters with simple teeth. A species closely resembling them, *Haplochromis bloyeti,* is found in many rivers near Lake Victoria. Quite possibly, almost all of the lake's species descended from this riverine species after it was introduced to the newly forming lake some 750 thousand years ago.

From the punctuational viewpoint, the critical fact is that there remain in Lake Victoria primitive forms resembling the ancestral riverine species. These have not been transformed into specialized animals, as a gradualist would imagine should have happened. A gradualist would, in fact, be hard pressed to explain why some lineages evolved at a remarkably rapid rate after becoming established within the lake, while their progenitors—the ancestral species of the lake—remain almost unchanged. The punctuational scheme, on the other hand, is quite compatible with these facts. There is a simple explanation. An average species of freshwater fishes evolves fairly slowly (our data suggest that it takes several million years for an average chronospecies of fishes to evolve sufficiently to be recognized as a new chronospecies). Thus, we would not expect species that first colonized the youthful Lake Victoria to have changed appreciably. Rather, according to the punctuational model, we would predict that if the ancestral species gave rise to strikingly new species, this would not

On the Rapid Origin of Species

have happened by their own transformation but by their budding off of distinctive new forms, which would, in turn, have budded off others. The result would be exactly what we see today: a mixture of the generalized ancestral forms and the specialized descendant forms.

The adaptive radiation of cichlid fishes in Lake Victoria magnificently illustrates two fundamental characteristics that I have already ascribed to adaptive radiation. One of these is that adaptive radiation typically follows the opening up of environmental space. In some instances this is achieved by extinction of a previously successful group of organisms, but in the case of Lake Victoria, it was accomplished by the very formation of the lake. Lake Victoria developed by the ponding of drainage by movements of the land associated with the rifting of the Earth's crust that has recently begun to disrupt the ancient continent of Africa. The unspecialized cichlids that first found their way into the lake, or into the ancestral bodies of water that became the lake, were confronted with few competitors or predators. Thus, there must have been little to obstruct their diversification into a variety of species specializing on various food resources and habitats within the lake. Here we have a strong hint of what is important for the occurrence of wholesale quantum speciation: ecological opportunity.

Given an opportunity like that of the cichlid fishes that were first washed into the incipient Lake Victoria, most groups of animals or plants would diversify rapidly, yet some groups, including the cichlids, display a tendency to undergo particularly spectacular radiations with great frequency. Not only Victoria, but other large African lakes, such as Tanganyika and Malawi, harbor huge cichlid faunas. I have singled out Victoria only because its youthful age points directly to quantum speciation. The cichlids of the tropical Americas have also radiated extensively. In a series of elegant experiments, Karel Liem of Harvard University has shown that, because they possess a complex pharyngeal jaw, the cichlid fishes are especially prone to evolutionary diversification of feeding habits. In effect, the rear portion of the jaw (the pharyngeal part) is decoupled from the front. By virtue of this arrangement, it is not required that the entire jaw system be engaged in both the gathering and the processing of food. Rather, the front part of the mouth and the front teeth can become specialized for gathering a certain kind of food, while the pharyngeal jaw can be specialized to act as a mill for grinding that particular food. For example, the pharyngeal jaw of cichlids that crush shelled mollusks bears heavy, round, molarlike teeth. At least nine species of African cichlids have remarkably specialized mouth parts for rasping scales from other fishes. Most of these eat almost nothing but scales. The two species of *Corematodus,* a genus endemic to Lake Malawi, resemble in form and color the species whose scales they steal. This evolutionary mimicry permits them to pass unrecognized among their potential victims in order to practice their unusual form of predation. Although consuming other items as well, one lantern-jawed species of African cichlid is specially adapted in a grisly fashion for tearing out the eyes of other fishes!

It is no accident that the several adaptive radiations I have described have taken place on islands, or they have occurred in lakes, which amount to islands of water in a sea of land. Insular settings of many kinds often come into being rapidly through volcanism, movements of the land, or the operation of other geological processes. If their boundaries are well defined, as is usually the case, their evolutionary products remain entrapped for us to study. Our frequent ability to pinpoint the recent time of origin of a discrete habitat such as an island or a lake may allow us to recognize that quantum speciation has taken place within it. How exactly is it that quantum speciation occurs on such a large scale within insular settings? A clue is provided by the early history of a population of small, silvery fishes in a twentieth-century body of water of southeastern California known as the Salton Sea.

Before 1905, the basin now occupied by the Salton Sea was a salt-encrusted depression inundated only intermittently by the Colorado River at flood stage. Between 1905 and 1907, the Colorado made a much larger contribution to the basin. Breaching an irrigation gap in its levee in the Imperial Valley, the river spilled into the basin, filling it to a depth of about sixty-seven feet. Thus formed the Salton Sea, which is really a lake whose waters happen to be saline, having dissolved some of the underlying salt deposits. The lake has shrunk to about two-thirds of its original areal extent, now measuring about ten by thirty miles, but is maintained at its present level by the periodic overflowing of nearby irrigation ditches.

The bairdiella is a small, marine fish native to the Gulf of California. In 1950 and 1951, sixty-seven bairdiellas were transplanted to the Salton Sea. The first successful spawning of the species in its new environment was achieved quickly, in 1952, but had bizarre results. It is estimated that between 13 and 23 percent of the young fishes resulting from the first spawning were visibly malformed in some way. Some were blind in one eye or displayed an abnormally small eye or pupil, some had strangely developed lower jaws or none at all, a few had snub-nosed heads, a small fraction displayed vertebral deformities, and a significant number possessed three anal spines rather than the usual two. During 1952 and 1953, these freakish traits persisted within

120

the population. By 1953, however, as the first-year class grew up and a new one was added, the population began to strain the food resources of the Salton Sea. As a result of crowding, far fewer of the abnormal juveniles that were born from the second spawning in 1953 grew large enough even to be observed. There was also increased mortality of the older freakish forms, so that by 1954 they constituted only 2 or 3 percent of the survivors of the original spawning.

The early history of the bairdiella in the Salton Sea suggests that much of the natural selection that occurs in nature is what we term stabilizing selection. That is, the selection tends to maintain the status quo by weeding out mala-dapted individuals. The first hatchlings of the Salton Sea emerged into a world that was essentially barren of competitors and predators. Until food supplies ran short and competition set in, the normal animal and the freak could flour-ish side by side. In other words, the Salton Sea experiment simply revealed how much potential variability we never see under normal environmental conditions.

We do not know what proportion of the abnormalities of the Salton Sea fishes were genetic in origin and therefore heritable. Possibly a large percentage were nongenetic errors of development that could not be passed on to off-spring, to take part in evolution. Still, we must imagine that some of the abnormalities were genetic, and for some small fraction of these we can envision a potential evolutionary role.

Let us imagine that a vacant body of water invaded by a small population of fishes happens to be somewhat larger and more complex than the Salton Sea. Let us further imagine that, because of divergent feeding preferences or other habits, different forms of fish within the population happen to favor different habitats within the large body of water. Possibly some small schools of closely related and therefore similar fishes become isolated by the flooding or drying up of certain areas. Such events, or what is called assortative mating (a tenden-cy for animals to breed with other animals of the population that resemble themselves) might then lead to the fixation of certain unusual traits within discrete populations. Fixation would occur over the course of several genera-tions, with natural selection guiding much of what happens from generation to generation. Once distinctive populations became reproductively isolated from others, they would represent new species.

The hypothetical sequence of events just described takes us into an area of biology where there is currently much controversy. One tenet of the Modern Synthesis has been that, in general, to evolve into a new species, a population must be geographically isolated from other populations of its species. Today, some workers suggest that complete isolation is unnecessary: a strong tendency for two different populations to occupy distinct, though adjacent, habitats may lead to the development of reproductive barriers between the populations, even though a small amount of gene flow may at first connect them (there may be some interbreeding at the outset). Thus, complete physical separation is claimed to be unnecessary. Another question is just what constitutes geographic separation. When we contemplate geography, we tend to think of distances measured in miles or kilometers. Large distances integrade with small distances, however. Is there any difference between being isolated from the rest of your species by a wide river or being isolated because you and your mate live your entire lives on a rock that is several yards from other rocks where other members of your species are found? If the limited mobility inherent to your pattern of behavior or body plan for many years prevents you and your progeny from breeding with populations inhabiting nearby rocks, you are effectively isolated, even though not on a geographic scale. For our purposes, the spatial details of speciation are not of great consequence—and they may, in fact, differ among the many varieties of animals and plants. What matters is that we have fossil evidence pointing to small populations as the kind that must accomplish quantum speciation. We have biological evidence that such speciation has indeed occurred in the recent past; and because our evidence here identifies insular areas as the Gardens of Eden, we may infer that the kind of release of variability observed for the insular Salton Sea bairdiella population may be of great importance. Newly formed insular habitats may allow so much previously constrained variability to emerge that rapid fixation of new features is rendered much more probable than under normal circumstances.

Since Darwin's analysis of evolution in the Galápagos archipelago, it has been understood that adaptive radiations are likely to occur in insular settings. The rapidity of such radiation has usually been underestimated, however. Also, it has commonly been claimed that insular radiations are atypical events that play little part in evolution in general. The idea here is that their products, though magnificent, are segregated from the rest of the world's inhabitants and doomed to extinction in isolation. To a geologist, such an argument holds little water. It is true that most oceanic islands and their biotas are condemned to a solitary life and an erosional death. Likewise, many lakes and their inhabitants are subject to a lonely demise by infilling and desiccation. Still, it is important to recognize that in identifying recent adaptive radiations we automatically single out islands and lakes because they are defined by strict boundaries. These boundaries may shift with the action of geological processes, but they remain intact for a time, retaining for our study the species that form within them.

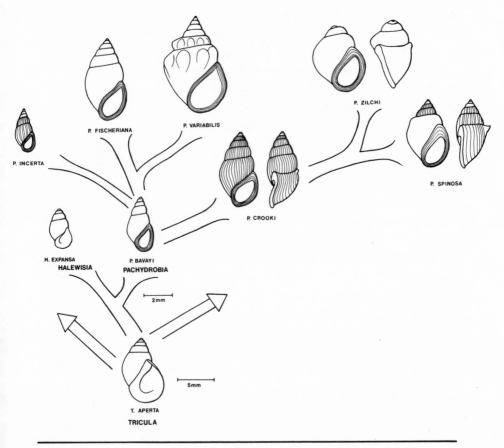

FIGURE 6.4

Hydrobioid snail species that have evolved as part of an enormous adaptive radiation in the recently formed Mekong River system of Southeast Asia. This diagram illustrates the apparent relationships among the species. The tribe of snails represented here includes three genera: *Halewisia, Pachydrobia,* and *Tricula.*

In other environments, insular settings must also form frequently but remain less long, releasing their biotic products to surrounding areas and leaving only vague indications of early boundaries. Continental regions provide the best general examples. For some of these we can find at least traces of evidence today.

Hydrobioids are small snails with coiled shells that inhabit both freshwater and marine environments. George M. Davis has uncovered an extraordinary adaptive radiation of these forms in the Mekong drainage system of Indochina. Here, approximately a hundred living species that did not exist until about ten million years ago are found in streams and rivers. Davis has traced the origins of the snail fauna to Gondwanaland, an enormous, ancient continent of the Southern Hemisphere. Tens of millions of years ago, Gondwanaland was fragmented to form the modern continents of Australia, Antarctica, South America, and Africa, as well as what is now peninsular India. The triangular peninsula of India moved northward as an island during the Cenozoic, and when it collided with Asia, the lofty Himalayas were formed. Riding aboard this moving island were hydrobioid snails, and when the island was attached to Asia, they made a landfall on the Asian continent. The uplift of the Himalayas that accompanied this continental collision led to the formation of the enormous Mekong River system that flows southward to the sea. In this manner, the stage was set for adaptive radiation: a new drainage system was created and simultaneously stocked with a small group of animals. Some ten million years of adaptive radiation since that time have yielded the large modern fauna of snails. Ironically, the uplift of the chain of mountains from the Himalayas to the Alps had the side effect of eliminating large bodies of water in the region of Yugoslavia, and here a large older group of hydrobioid snails was driven to extinction by loss of habitat.

Sagas like that of the Mekong snails can be reconstructed for terrestrial life as well. Nearly everyone is familiar with the strange marsupial mammals of Australia, such as the kangaroo, the koala, and their diverse pouched relatives. Aside from species introduced by humans, there are no advanced placental (nonmarsupial) land mammals on the enormous island-continent of Australia. While small mammals were well established elsewhere in the world during the Mesozoic Era, it was not until the Cenozoic Era that the first mammals reached Australia, apparently from South America by way of Antarctica, which at the time was positioned so as to form a kind of continental bridge. These early mammalian pilgrims happened to be marsupials, and their ensuing adaptive radiation took place within a vast insular region. Instead of yielding antelopes or horses, evolution here happened to produce the curious bounding animals that we call kangaroos. The Earth's climate has undergone a cooling trend during the marsupial radiation. Much of the fauna of Australia would have died out as a consequence, had not the continent simultaneously been moving toward the equator. Possibly its marsupial faunas will live out an odyssey that at some future date will have them spilling onto a more northerly continent, such as Africa or Asia, with which Australia may eventually collide.

We have evidence of a comparable event in our own hemisphere. Throughout most of the Cenozoic Era, South America stood in solitude as a large island

FIGURE 6.5

Some of the animals produced by the South American adaptive radiation of marsupials that occurred when South America was an island continent. *Coenolestes* (A) resembles a rodent. *Thylacosmilus* (B) was closely similar to a sabertooth cat. *Argyrolagus* (C) was convergent in form with a kangaroo rat. *Lutreolina* (D) belonged to the opposum family. *Necrolestes* (E) was a burrowing animal with an upturned snout. *Borhyaena* and *Prothylacynus* (F and G) were catlike carnivores.

continent, unattached to North America. Here primitive marsupials, such as the ones that made their way to Australia, underwent their own adaptive radiation, along with the radiation of a few nonmarsupial mammal groups that somehow reached South America from the north. (The recently extinct giant ground sloth and giant armadillo that Darwin uncovered in the fossil state were unusual products of the limited South American radiation of nonmarsupial mammals.) Then, very recently, between three and four million years ago, there occurred a remarkable event in the history of mammalian geography. The Isthmus of Panama was formed by regional elevation (recall Darwin's experience with Andean earthquakes and uplift). The result was a corridor across which faunas could migrate north or south. As it turned out, most of the migration was a southward movement of placental mammals from Central to South

America, but some mammals, such as the familiar opossum (a marsupial) and the armadillo, made their way north.

It seems evident that we must regard insular adaptive radiations as the source of much evolutionary change. It is true that some of these events—for example those of permanently isolated and geologically evanescent settings such as lakes and oceanic islands—may be evolutionary cul-de-sacs, but we must not be misled by these. They stand out because of an observational bias. River systems, confined terrestrial habits, and many other insular geographic settings may remain isolated long enough to foster adaptive radiation and yet soon have their boundaries break down. Rivers capture other rivers, mountain chains wear down, continents collide, and climates fluctuate. Such vicissitudes create corridors for migration. Insular radiations must continually take place, but their individual histories are quickly obscured. Even for many of the most recent examples, rupturing of boundaries and leakage of biotic productions has left us unable to reconstruct details or even recognize the insular nature of radiations.

The scale of insular radiation ranges from the production of a few species in a local area to colonizations of global proportions. With the demise of the dinosaurs, the continents of the world were automatically transformed into vast depauperate islands available for renewed occupation by terrestrial vertebrate life. Is it any wonder that, as I have described, most of the modern orders of mammals had appeared by the time the ensuing Age of Mammals was only about twelve million years old? During this interval, spectacular insular radiation proceeded on a continental scale.

In contemplating large, discrete adaptive radiations that may indeed account for most evolution, we must not lose sight of the fact that opportunities for speciation are widespread. Environments are patchy and remain in a state of flux. Quantum speciation is continually occurring here and there throughout the world, often not as a part of insular radiation but simply in the form of solitary local events.

As we have seen, the genetic underpinning of the Modern Synthesis placed emphasis on the gradual restructuring of large species by the piecemeal retention of favorable mutations. But where do we turn for a genetic understanding of rapid transformation within small populations? We have no simple answer. In fact, in 1974 the prominent geneticist, R. C. Lewontin confessed, *"we know virtually nothing about the genetic changes that occur in species formation"* [the italics are his].[3] It is nonetheless gratifying that new discoveries are leading in fruitful directions. We are learning to appreciate that small genetic changes can engender great changes in bodily form. As with the evidence for quantum speciation, this genetic information is of great value, not because it confirms the punctuational model (here, we must rely heavily on fossil data) but because it suggests a feasible mechanisms. The fossil record is not demanding the impossible.

One of the startling new facts of genetics is that many of the proteins that form cells and operate them have great similarity throughout the animal world. It is estimated that 98 or 99 percent of the protein structures of humans and chimpanzees are the same! Clearly, evolution is reshaping animals in major ways without drastically remodeling the genetic code. During the past few years, geneticists have begun to ascribe great importance to entities that are often loosely termed regulatory genes. These are genes that play some role in the switching on and off of structural genes. Structural genes are the worker genes, the ones that represent templates for the formation of the proteins that form physical structures and that mediate chemical reactions within the cells. As our bodies change from hour to hour and from year to year, different parts of our genetic message come into operation—different structural genes are at work forming proteins. What is important about regulatory genes is not simply that, by the operation of hormones and other agencies, they switch structural genes on and off as needed. What is important is that the mutation of a single regulatory gene can have profound heritable effects. Some regulatory genes command many structural genes. The mutation of one of these high-ranking regulators can eliminate or redirect a major aspect of growth or development. Precisely how regulatory genes operate is a matter of current research. Answers are not easily divined because many regulatory genes do not produce proteins or other compounds that are easily identified and studied.

The classical example of a major developmental change that may be controlled by a single gene is metamorphosis in amphibians. Many amphibians begin life as tadpolelike infants that hatch in water and much later undergo metamorphosis into four-legged terrestrial animals. The Mexican axolotl is famous in biological circles as a salamander that never grows up. Its pituitary gland simply does not adequately stimulate the thyroid gland with a hormone. As a result, development fails to proceed beyond the early, aquatic stage. Because this type of arrested development characterizes the axolotl, this animal reproduces in the "juvenile" state. What is striking from an evolutionary standpoint is that injection of the missing hormone induces the axolotl to mature fully and take up an adult existence on the land. Without a doubt, this was the way of its ancestors. It is the way of most salamanders. The axolotl is probably but a single genetic step from an ancestor of this type—it takes only a

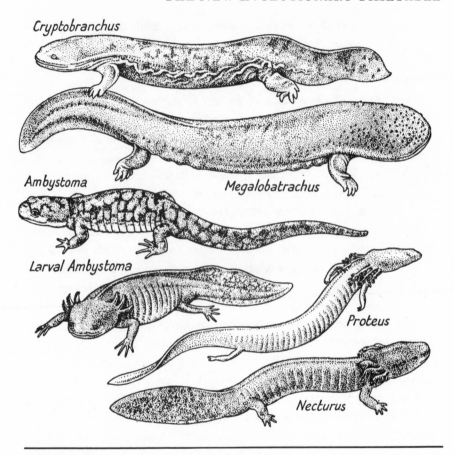

FIGURE 6.6

"Juvenilized" amphibians that, because of simple genetic changes, remain in the water throughout life instead of metamorphosing to become terrestrial salamanders. One adult terrestrial salamander, *Ambystoma*, is figured, along with its larva, whose tail fin for swimming and external gills resemble those of the "juvenilized" forms.

single gene to activate a hormone—but has a totally different adult mode of life. The kind of genetic change that has produced the axolotl is by no means unique. This species is the most famous amphibian that never grows up, but it is not the only one.

The profound influence that a few genes can exert is illustrated by the origin of the giant panda. For years it was debated whether the giant panda and lesser panda belonged in the bear family or the raccoon family. In 1964, Dwight Davis published a definitive anatomical monograph on the giant panda, in which he showed that this animal is descended from the bears, while the lesser

On the Rapid Origin of Species

panda has instead evolved from the raccoons. The similarity of the two pandas is the result of convergence, or evolution that has moved the two groups from different ancestral anatomies toward the same body plan. The giant panda is essentially a machine for eating bamboo. It spends much of its time simply sitting and chewing. The teeth and jaws are enlarged and strengthened so that they effectively process the coarse food. In complementary fashion, the rear portions of the animal are weakly developed. The genitalia and skeletal features of the pelvic region are aberrant in form. Davis showed that most of these features are essentially the result of a simple evolutionary change in growth gradients within a bear, so that the front of the animal became especially well developed at the expense of the rear portion. Any animal has fixed amounts of energy and materials available for growth, and in the panda these are directed preferentially toward the head region. Two salient points emerge from this analysis. One is that the essential change from bear to panda could have been achieved by a very small number of regulatory genetic changes—perhaps just two or three. We may infer that the evolutionary transition could have been quite sudden. The second point is that at least some of the unusual features of the rear portion of the panda were not themselves selected for, but rather, they seem to have been the automatic consequences of selection for a strengthened head region. Interestingly, pelvic abnormalities similar to those of the panda are found in the bulldog, which was bred artificially to have a massive biting apparatus for fighting and which ended up with the weak and reduced hind quarters that in cartoons are often rendered as ludicrously undersized. In any animal, many genes control more than one feature of anatomy, and this inevitably results in evolutionary compromise. In both the bulldog and the panda, the sacrifice of strength in the hind quarters has been tolerable. The bulldog is well cared for in domesticity and is not required to chase down prey. The panda is a clumsy teddy bearlike creature, but having few natural enemies and feeding mainly on bamboo, it is seldom required to flee from danger or pursue agile prey.

The giant panda is so distinctive that it is placed in a family or subfamily of mammals in which there is no other living species. The fossil record of the pandas is in China and is not well known, but it is possible that the peculiar family or subfamily that we call pandas emerged by way of a single speciation event. We can imagine that somewhere in a local bamboo forest high in the Himalayas, during a brief interval of geological time, a very small population of bears underwent the basic transition.

Other major evolutionary changes in animal form may also be viewed as alterations in the relative growth rates of different parts of the body. As in the origins of the panda and axolotl, such restructuring may result from simple changes in gene regulation. The genetic command to grow at a given time, at a given rate, and in a given way is easily stifled or modified, and the result may be dramatic.

Quite apart from what the fossil record says, there are biological reasons for believing that major restructuring of an animal or plant is most likely to take place within a small population. One consideration is that, if natural selection is to be responsible for fixing the new features throughout a population, the selection must operate consistently. If the new features are not of value nearly everywhere, they will not spread throughout. A large population, such as that constituting a typical full-fledged species, will normally occupy many different patches of environment. Each patch will support its segment of the total population in a unique ecological context, with a particular set of predators, competitors, food resources, and physical habitats. How can we expect a dramatically new shape or behavioral pattern to be favored consistently in nearly all of the varied patches of environment? Herein lies one of the problems for evolution within large populations. The peppered moth of Great Britain is famous for having become black, rather than speckled gray, in industrial cities during the Industrial Revolution. Here selection has obviously been at work; where soot is heavy, the black color is relatively inconspicuous to predatory birds. The truth is, however, that only some populations ever became dominantly black. In unpolluted areas, most of the moths have remained a speckled gray color, and even in polluted areas not all moths are black. This celebrated example of evolution within living memory is indeed a good example of selection, but not of the transformation of an entire species.

A second problem for the rapid restructuring of large populations is gene flow. There is much evidence that the patch distributions of many species result in very little movement of individuals, spores, or seeds from patch to patch throughout the species' range. Thus, if a dramatic new adaptation crops up in one area, it will be unlikely to spread throughout the entire range of a species, even if it has the potential to displace preexisting adaptations.

Still another reason for giving attention to small populations is that the fixation of strikingly new adaptations often seems to require inbreeding. One factor here is the tendency for some important new genes to be recessive in their original genetic context. They will only have physical expression if two animals that possess them breed to produce homozygous offspring (recall the earlier discussion of red and white flowered peas). Another hindrance appears to be

On the Rapid Origin of Species

that some unusual individuals that might initiate rapid evolutionary divergence are genetically, physically, or behaviorally incapable of producing fertile offspring, except by mating with their own kind. An important factor here may be that chromosomal changes play a role in the initiation of major evolutionary steps. There is much evidence that chromosomal rearrangements occur in association with speciation events. This we can see quite simply from the fact that many closely related species are characterized by distinctive chromosomal features. Chromosomal mutations or rearrangements may come about by accident during the cell divisions that produce sex cells; here chromosomes divide, and in dividing they may be permanently fragmented, or segments of them may be reversed or displaced. The genes on the chromosomes may not be altered in this process, but the way that they function may be affected. It has been suggested that the mode of operation of a regulatory gene relates to its position relative to the genes that it controls. If this is true, then it is easy to see how chromosomal rearrangements may have profound evolutionary effects. Imagine, for example, that a regulatory gene from the chromosomal neighborhood of one or more of these structural genes and into the vicinity of others may have profound effects. It may cause changes in the sequence of development of an animal or plant, or it may change the rate at which different organs grow.

Even if these "position effects" of genes are not of great adaptive importance, chromosomal variations among species may still be telling us something of great importance. They may represent markers, showing that most major evolutionary transitions occur within small populations, as so much other evidence seems to indicate. The explanation is simple. Individuals whose parents had differing chromosomal arrangements tend to be maladapted; they are the products of a bad match. Thus, while for any new genetic feature to spread throughout the range of a large species is difficult, for a chromosomal rearrangement it is nearly impossible. Fixation of a new arrangement is usually likely only in a small population, where inbreeding can rapidly allow it to become established in a happily paired state.

It is not simply many species that have unique chromosomal configurations, but many genera and families. This striking piece of information suggests that many of these distinctive larger groups can be traced back to single ancestral species that formed by way of small, inbreeding populations. We do not know for sure, however, that the first species with a new chromosomal pattern possessed most of the adaptive features that now distinguish the genus or family in which the chromosomal pattern is displayed.

It is clear that many new adaptive features do not emerge at the precise time when speciation occurs. In truth, we cannot pinpoint time of speciation because there is no way of determining just when a diverging population can no longer breed with its parent species. Hybridization, or the interbreeding of differing populations, may remain possible for some time, although hybrids are often infertile or weakly fertile. Polymorphism, or the occurrence of varied forms within a species, sometimes develops to a remarkable degree. A new shape is likely to play a role in phylogeny, however, only if it is fixed within a discrete species that can persist and itself bud off additional species of similar form. In other words, speciation is important even though it may follow the emergence of a new adaptation.

Thus, there is no need for the punctuational model to focus strictly upon the moment of speciation, which cannot, in fact, be identified. The small populations in which rapid changes tend to occur may represent incipient species, already reproductively isolated, or they may for some time remain interfertile with other populations of their parent species. The problem of recognizing species seems to be much greater in the study of fishes than in the study of large mammals. It seems also to be much greater in the study of corals that build reefs in tropical seas than in the study of sea urchins, some of which inhabit these reefs. Modes of speciation may, in fact, differ significantly from group to group. The important point for the punctuational scheme is that there are discontinuities of form and behavior between many related species and between many related populations and that these discontinuities commonly arise by the rapid divergence of small populations.

If gene flow within a large species is weak, then the entire species is likely to evolve significantly in a particular direction only if two conditions are met. First, natural selection must be consistent in space and time: throughout the species' range there must be a tendency for a particular kind of selection to be sustained. Second, everywhere within the population appropriate variability must be available upon which this particular kind of selection can operate. For unusual genetic changes, the second criterion is unlikely to be met. How probable can it be that there will crop up in nearly all populations of a species' range a particular pattern of chromosomal reorganization or a rare mutation of a particular regulatory gene?

For some adaptive changes, the problem is not so great. For example, it is obvious that in nearly all abundant species there is substantial variation in body size. Some of this variation is certainly due to inheritance, as we have all seen by observing the history of human families. Many species originate at small body size, yet there are often advantages for their members to become larger.

On the Rapid Origin of Species

Large animals may, for example, capture food more effectively, ward off predators more easily, or win more battles for mates. We can therefore predict that increase in size should be one of the most common kinds of evolutionary change within established species. Our prediction is borne out by the evidence. Cope's Rule, a venerable dictum based on abundant fossil data, states that most species tend to evolve toward larger body size. Of course, average body size of a species will not increase beyond the optimum condition for a given time and place, even though variability in size will continue to be generated.

We do not know exactly what evolutionary changes other than increase in body size are most likely to take place throughout a populous species. Presumably such changes should be ones for which variability is readily available and ones which are likely to be of value to all populations, regardless of peculiarities of the environment. Possibly many such changes relate to reproductive output, which is of course an important component of selection. We might imagine that sexual selection, in which certain kinds of animals are favored as mates, might be a relatively potent source of widespread change. If, for example, mating behavior in a species centers around a particular visual attractant, such as a peacock's tail, then throughout the species' range selection will favor those males whose tails are particularly well developed.

The foregoing examples suggest caution. In focusing on small populations, we must not ignore altogether certain modifications of full-fledged species. Let us, however, return to rapid changes within small populations in order to assess more fully why in the past they have been accorded so little importance.

Historically, the most common objection to the idea of rapid but extensive evolutionary change within small breeding groups has related to numerical probability. It has simply been judged highly unlikely that a strange new kind of individual would ever find a mate. This viewpoint, which has most commonly been expressed by biologists rather than by paleobiologists, reflects a lack of appreciation of the vastness of geological time. The origin of the snails, which constitute a large class of mollusks, serves to illustrate this point.

Snails are characterized by a peculiar anatomical ground plan, which involves a twisting of the elongate snail body through a half a circle so that the anus lies nearly above the head. This twisting, which is a separate feature from the coiling of many snail shells, is called torsion. Torsion occurs by the differential development of two muscles in the juvenile snail. It is remarkable that in some primitive snail species the twisting process requires only a few minutes. So simple is the mechanism of torsion, despite the profound result, that it has long been held likely by some experts that torsion is the result of a single mutation.

This idea gains strong support from the fact that snails that have evolved in such a way as to lose torsion have done so irregularly, with different features of anatomy becoming untwisted to different degrees. Obviously detorsion has evolved by progressive steps. Had torsion itself evolved in this way, rather than as a single step, we would expect to see it, like detorsion, developed to different degrees among different snails. This is not the case.

Snails evolved at some time during the Cambrian, the first period of the Phanerozoic. Let us cast our thoughts backward more than 500 million years, to an even earlier Cambrian world in which no mollusk was torted. Is it unlikely that an individual as strange as the first torted snail would then have found a mate among its contemporaries? Let us assume that untorted "presnails" were crawling over the sea floor for some ten million years prior to the big event. Let us further suppose that during this interval there existed a worldwide total of a thousand species of "presnails." Finally, let us assume that at any time an average species was represented by ten million females. All of these numbers are reasonable, if only approximate. What they tell us is that if every female produced a brood of offspring annually, there would altogether have been something like a hundred thousand trillion broods of presnails in which torsion might have become fixed. This number is so astronomical that there remains absolutely no foundation for the claim that a feature like torsion would have had little chance to arise by a single mutation and quickly become established. One likely mechanism would have been for a female snail to have experienced a mutation of her gonad tissue so that some or all of her offspring happened to be torted. Then, inbreeding among siblings could have fixed the strange new condition. This might have occurred because two or more torted brothers and sisters happened to become spatially isolated, or it might have occurred because a torted snail could only have bred successfully with another torted snail. The second condition would immediately have conferred reproductive isolation. Still another possibility is that only a single torted individual was born, but that it was able to interbreed with other members of its species, and spatial separation and inbreeding then fixed the torted condition. Regardless of the precise mechanism, the rapid establishment of torsion gave birth to a new class of animals, albeit a class that, aside from its torted body, was not very different from the class from which it evolved. Torsion seems to have turned out to have value in bringing the sense organs forward, to a position above the head, and in permitting the subsequent evolution of an operculum or trapdoor that protects the snail when it withdraws into its shell.

Another way that we are able to appreciate the vast opportunities for quantum speciation is to observe the population structures of certain living species. *Clarkia* is an interesting genus of annual plants that has been studied in California by Harlan Lewis. Lewis has written:

> Deviant marginal populations are so frequent in *Clarkia* that one often stumbles onto them without making a deliberate or systematic search. They may differ from the parental species in any one or more ways, including morphology, genetic compatability, breeding structure, and chromosome arrangement. They all have, however, two features in common; they are ecologically marginal, and the discontinuity with the parental populations is abrupt. One may assume that most deviant marginal populations are ephemeral and destined to become extinct in situ. Few warrant taxonomic recognition, but some have persisted and flourished as distinct species, and occasionally, we may suppose, a deviant may set the stage for a successful new phylad of species [a new genus, for example]. . . .[4]

Clarkia may be an exceptional genus in the frequency with which its species generate distinctive small populations. But given vast stretches of geological time, other species have many opportunities for quantum speciation, which is, after all, a rare event.

As I have noted in an earlier chapter, the prominent experimental geneticist Richard Goldschmidt was ostracized from the Modern Synthesis for his radically punctuational views. His ideas on large-scale evolution were widely derided. In retrospect, some of Goldschmidt's views still appear extreme, yet many were no farther from the punctuational model I have been describing than the prevailing views of the Modern Synthesis were in the opposite direction. Goldschmidt erred in viewing only the whole chromosome, and never the gene, as the unit of selection. He also exaggerated the importance of chromosomal rearrangements, viewing them as the only mechanism by which a species could form. Even so, he prophetically focused attention upon chromosomal change, on the role of single mutations in effecting rapid changes in growth gradients or developmental sequences, and on what we now call quantum speciation. Goldschmidt's most controversial construct was the "hopeful monster," the single animal supposed to constitute a new genus or family at birth. Otto Schindewolf, a German paleontologist, was driven to similar views by the antigradualistic evidence of the fossil record. As I have already noted, Schindewolf envisioned the first bird hatching fully formed from a reptile egg! These men were given no quarter, however, and quantum speciation of any sort was rejected along with their extreme views. In retrospect, it seems to me that Goldschmidt deserves posthumous accolades for his steps in the right direction, though they may have been steps too far.

At the end of chapter 3, I offered several reasons for Darwin's earlier rejection of rapid transition within small populations, and to these we may now add another. It relates to the importance of inbreeding to quantum speciation. Certainly, as nearly everyone knows, long-term inbreeding can be deleterious, but in the fixation of new features, several generations of inbreeding need not spell disaster. There is chromosomal and other evidence that many hale and hearty species have descended from the offspring of a single female, and this implies an early history of inbreeding. Darwin, however, saw inbreeding as anathema to the well-being of any population. This was the result of his experience with domestic "productions," as he called them. It was partly because of this exaggerated view that he granted no evolutionary role whatever to "sports" of nature—monstrous or visibly deviant individuals.

In his unpublished "Essay of 1844," Darwin had actually attributed modest evolutionary significance to "sports," writing:

> So in the state of nature some small modifications, apparently beautifully adapted to certain ends, may perhaps be produced from the accidents of the reproductive system, and be at once propagated without long-continued selection of small deviations towards that structure.[5]

This attitude soon changed, however. In the first edition of the *Origin*, Darwin ignored the possibility of such rapid changes, and, when he went to press with the third edition, he had shifted to a totally negative view:

> It may be doubted whether sudden and considerable deviations of structure such as we occasionally see in our domestic productions, more especially with plants, are ever permanently propagated in a state of nature. . . . They would, also, during the first and succeeding generations cross with the ordinary form, and thus their abnormal character would almost inevitably be lost.[6]

Here Darwin fell victim to the contemporary concept of blending inheritance. He concluded that "sports," if they could reproduce at all, would have their unusual traits progressively diluted into oblivion as generation followed generation. I have noted previously that a resort to inbreeding within small populations of deviants would have represented an escape from this apparent problem, yet Darwin eschewed this alternative. His apparent reason was his conviction that inbreeding is always deleterious (as it may indeed be, if carried on for many generations). Chapter Seventeen of his book *The Variation of Animals and Plants under Domestication* was descriptively, if awkwardly, entitled "On the Good Effects of Crossing, and On the Evil Effects of Close

Interbreeding." Here he reported that close inbreeding was "generally believed" to result in "loss of size, constitutional rigour, and fertility, sometimes accompanied by a tendency to malformation."

Darwin's exaggeration of the evils of inbreeding was a prime reason for his refusal to grant a significant evolutionary role to small populations. Conversely, the apparent need for outbreeding impelled Darwin to focus upon very slow evolution within large, established species. It was partly for these reasons that Darwin, as Huxley noted, heaped unnecessary problems upon himself and upon natural selection by clinging to the time-worn cliché: *Natura non facit saltum.*

CHAPTER

7

Human Origins

NOWHERE is the gradualistic tradition more evident than in our study of ourselves. In his book *Mankind Evolving* (1962), Theodosius Dobzhansky wrote:

> The evidence now available . . . is compatible with the assumption that, at least above the australopithecine level, there always existed only a single prehuman and, later, human species (which evolved with time from *Homo erectus* to *Homo sapiens*). Mankind was and is a single, inclusive Mendelian population and is endowed with a single, corporate genotype, a single gene pool.[1]

Dobzhansky allowed for some evolutionary branching earlier in our ancestry, but maintained that the general hallmark of evolution within the human family has been the gradual transformation of species. Dobzhansky was not alone in this opinion. Among physical anthropologists and paleontologists specializing in the study of human origins, the convention has been to reconstruct our pedigree in terms of the gradual humanization of an apelike ancestor. In fact, until recently, a popular tenet in this field has been the single species hypothesis, which suggests that competitive interactions would have prevented two or more humanoid species from ever existing simultaneously. The idea was simply that the world is not big enough for two of us. Even today, the most common interpretation of human evolution places us at the end of a single lineage tracing back to a slender australopithecine—a somewhat apelike humanoid that I will describe shortly. This view has prevailed to the point that it has become difficult to extract objective observations from many writings on the subject.

Human Origins

Fossils have often been described in terms of how they seem to fit a preconceived gradualistic scheme rather than in terms of their fundamental attributes.

In taking up the subject of human origins, let me stress an important point. In suggesting a punctuational pattern for human evolution, I will not simply be extrapolating from what we know of other animal groups. The main argument I have directed against gradualism—the argument that established species tend to last for long geological intervals without evolving sufficiently to change names—holds for known humanoid species. In fact, it is a great irony that, while gradualism still prevails in the study of human evolution, now the evidence here seems powerfully in favor of the punctuational model. It appears that our own species, in particular, is the product of a remarkable event of quantum speciation.

The failure of a punctuational scheme of human evolution to surface until recently can be explained in part by the prevailing climate of gradualism within evolutionary science in general. For decades, the availability of only scattered fossil remains of the human family permitted a gradualistic orientation to dominate within physical anthropology, a field which has itself generated rather little basic evolutionary theory. In recent years, the factual picture has changed.

With enthusiasm for the single species hypothesis already flagging, Richard Leakey and Alan Walker[2] in 1976 described remains of *Homo erectus,* a species of our own genus, found with a species of robust australopithecine within a thin stratigraphic interval in East Africa. The two species clearly coexisted early in the Ice Age, from before 1.6 million years ago to nearly 1.3 million years ago. This unequivocal evidence closed the door on the single species hypothesis and opened the way for punctuational interpretations. It showed that branching of lineages has been a fundamental feature of hominid evolution.

While branching is required in the punctuational model, the mere presence of branching does not validate the model. Even with some branching, most human evolution could still have taken place within established lineages; branching might not have represented quantum speciation, but only gradual divergence. This point, and the tradition of extreme gradualistic thinking, is reflected in the work of Louis Leakey, who was an arch opponent of the single species hypothesis. Leakey believed that our species, *Homo sapiens,* overlapped in time with the extinct species *Homo erectus,* but he remained a staunch gradualist. As a result, he claimed that we could not have evolved from *Homo erectus,* and he looked to another lineage for our ancestry. His assumption was that we could only have evolved by the complete transformation of another species; we could not have branched suddenly from *Homo erectus.*

What now seems to imply a dominant role for rapidly divergent speciation in human evolution is the great temporal longevity of fossil species. Somehow modern humans have evolved from australopithecine antecedents during the past two or three million years. This fact must be reconciled with the evidence that, meanwhile, certain hominid species have persisted with little change for approximately a million years!

To the degree possible, I will provide a profile of each of the fossil hominid species that lived during the past four million years. It will be evident that, owing to the fragmentary nature of the evidence, the classification of many fossils remains controversial. The telling point, however, remains. Certain species in our ancestry are represented by populations that attest to an enormous time span over which little evolutionary change took place. We will begin the central story in the Pliocene, more than three million years ago, with the oldest known australopithecines, but first it will be useful to step back a bit farther into the past, in order to catch a glimpse of our ancestors coming down out of the trees.

In his book *The Descent of Man* (1871), Darwin concluded that our species is descended from a "hairy, tailed quadruped, probably arboreal in its habits."[3] Darwin's deduction was well founded. The remains of one such creature—a monkeylike animal with apelike teeth—were described by Elwyn Simons in 1962, shortly after their discovery in thirty-million-year-old Oligocene deposits in Egypt. The location of the find was incorporated into the generic name chosen for this early ape or apelike monkey: *Aegyptopithecus*. *Aegyptopithecus* was a tailed animal about the size of a house cat. The structure of its limb bones suggests that it traversed tree branches on all fours. Its apelike dentition included large canines (eye-teeth) in male animals only, and this disparity between genders possibly relates to a way of life in which males defended complex social groups. Advanced social organization is compatible with the large brain volume of *Aegyptopithecus*—about thirty cubic centimeters, which in light of the animal's small size may have made it the smartest animal in its Oligocene world. *Aegyptopithecus* looks indeed like a good ancestor for apes and humans—like the animal that Darwin prophetically envisioned—although we do not know for sure that it occupied such an evolutionary position.

During the latter part of the Oligocene and throughout most of the Miocene, for a total of perhaps twenty million years, primitive apes known as the dryopithecines underwent a sizable adaptive radiation in Africa and Eurasia. These animals were apparently derived from *Aegyptopithecus* or a close relative, and they, in turn, gave rise to modern apes and also to the family of man. The

FIGURE 7.1
Reconstruction of *Aegyptopithecus,* a thirty-million-year-old animal of the type that may have been ancestral to apes and humans.

time of divergence of modern apes and hominids is still in doubt. Dryopithecines themselves, as represented by their most familiar representative "proconsul," has postcranial skeletons reminiscent of *Aegyptopithecus.* They were apparently also primarily four-legged branch walkers with short forelimbs, but there are indications that, like modern chimpanzees, they occasionally swung from branches and also walked on the ground.

The Miocene Epoch, lasting from about twenty-two to five million years ago, was a time of great success for apelike animals. Today, in comparison, apes are very restricted in geographic range and number of species. Monkeys are far more successful. It is the hominid line of descent from dryopithecines that concerns us, however, and here the central figure may have been *Ramapithecus,* commonly recognized as the first hominid. *Ramapithecus* lived in Africa and Eurasia during the interval from about fourteen to eight million years ago. While the dryopithecine apes retained large canine teeth and relatively small molars, *Ramapithecus* exhibited a very different pattern, which appears to place it in an intermediate position between dryopithecines and more modern members of the human family: its canine teeth are reduced in size, and its molars are enlarged. This earliest genus of hominids was apparently well adapted for processing plant materials. At present we possess few postcranial bones of *Ramapithecus,* and we are unable to determine whether or not its species had left the trees for a life on the ground.

As I will outline, we have conclusive evidence that the succeeding group of hominids, the australopithecines, were two-legged ground walkers. Unfortunately, the latest Miocene and very earliest Pliocene (the period from about eight to four million years ago) has revealed little of the assumed transition from *Ramapithecus* to the australopithecines. Nonmarine sedimentary deposits of this interval are not well developed in many areas of Africa and Eurasia and remain poorly studied. Australopithecines younger than four million years have become rather well known, however, and so have members of the human genus, *Homo.* We will now examine the commonly recognized species of these groups—species which shed much light on the pattern of human origins.

Australopithecus africanus has often been regarded as the direct ancestor of the human genus, *Homo.* Its existence spanned the interval from three million years ago to two million years ago, but the species may well have come into existence earlier and survived to perhaps 1.6 million years ago in East Africa. In other words, *Australopithecus africanus* probably existed for at least a million years and may have persisted much longer. It was a small, slight animal, probably under four feet tall and weighing less than fifty pounds. We can refer to it informally as a slender australopithecine. A particularly human aspect of its skeleton was its dentition. Here it shared with our species several traits, including reduced canines and curved rather than straight rows of cheek teeth. The brain capacity of the slender australopithecines averaged slightly less than 500 cubic centimeters, approximating that of a gorilla. Much of any animal's brain is taken up with the control of musculature, and brain size therefore tends to increase with body size (though not as rapidly—an increase in body size by a given percentage requires a smaller increment in brain size). Because a gorilla is a much larger animal than *Australopithecus africanus,* we may infer that *Australopithecus africanus* may have been smarter. Perhaps more of its nearly equal-sized brain was given over to thinking. The slender australopithecine must have lived more by wits than by brawn.

The skull of *Australopithecus africanus* differed from that of humans in more ways than its smaller brain capacity. The cheek bones of the slender australopithecine projected, flangelike, from the sides of the skull, and its very long face sloped forward to a larger jaw, though the chin was much less prominent than ours. *Australopithecus africanus,* like our species, was a fully upright creature. This we can see from various skeletal features, but most especially

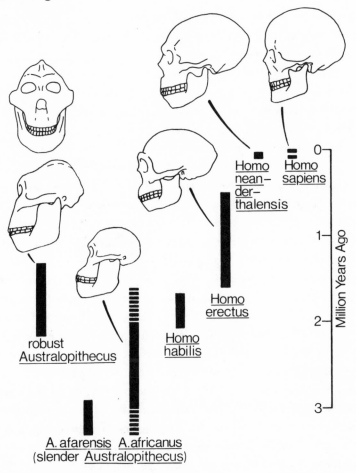

FIGURE 7.2

Geological durations of recognized species of the human family. Dotted segments of bars indicate uncertain occurrence.

from its basketlike pelvis—a pelvis built along human lines to support the body above the legs. An ape's pelvis is more elongate, extending for some distance along the lower back; its shape reflects the absence of fully upright posture. The configuration of *Australopithecus africanus* sheds light on an important question of human evolution. This animal's anatomy reveals that an upright posture evolved long before brain development approached a modern human level.

Australopithecus africanus is found in association with animals of the sort that roam over grasslands and savannahs. Apparently it lived in the open rather than in forests, and it seems to have been a generalized feeder. We do not know whether it was a big game hunter. Certainly there is no evidence whatever to support Robert Ardrey's idle assertion that our African genesis was from ancestors that were vicious carnivores.

Australopithecus robustus is a second australopithecine that bears sufficient resemblance to the species just described that many workers assign it to the same genus, although some separate it as the genus *Paranthropus*. Whatever generic assignment is favored, the second part of the name, *robustus,* is descriptive, at least with regard to the animal's head. The cheek bones are even more prominent, flangelike projections than in *Australopithecus africanus*. Not only are the molars quite massive, but the premolars are large and molarlike. In contrast, the canines are much reduced in relative size. The tooth structure of the robust australopithecine suggests a vegetarian diet. The massive molars were set in heavy jaws, and a sagittal crest, or ridge, ran down the middle of the skull for attachment of powerful jaw muscles. Big-brained animals often have such large skulls that no attachment crest is needed, but the robust australopithecine had a brain that was similar in size to that of a gorilla, which also has a sagittal crest. The face of *Australopithecus robustus* was long and slightly sloping, like the face of its slender cousin. Although it was apparently a more heavily built animal than *"africanus"* and possessed larger teeth and jaws than our own, *"robustus"* was probably still smaller in stature than a human, to say nothing of a gorilla. As we shall see, our fossils of slender australopithecines may actually represent more than a single lineage. The same is true of our fossils of robust forms, which display some variation in shape. Nonetheless, the story is similar with regard to rate of evolution. However many lineages of robust australopithecines may have existed, they apparently did not differ greatly from each other in form, nor did they evolve markedly. It is universally agreed that robust australopithecines represented a specialized offshoot of hominid evolution that had no place in the ancestry of our own species. Even so, two observations are relevant to our story. First, these herbivores lived in Africa for approximately a million years (from perhaps 2.3 to 1.3 million years ago). Second, for several hundred thousand years, they appear to have lived side-by-side with their slender cousin, *Australopithecus africanus,* and for a brief interval, at least, they coexisted with *Homo erectus.*

Homo sapiens might seem to need no introduction, but its shape must be contrasted to the shapes of less familiar hominid species. The australopithecines represent where we are coming from in reconstructing human origins. *Homo sapiens* represents where we must go.

Human Origins

The modern human brain averages about 1,330 cubic centimeters, compared to the 500 or so for australopithecines. On the average, we are a foot or more taller than the slender australopithecines, and correspondingly heavier, but this can account for only part of the difference in brain size. The brain volume of a single individual must not be taken too seriously as an indication of the intelligence of that individual or its species. Brain volume varies greatly within a species (for example, from 1,000 to 2,000 cubic centimeters in modern humans) without close correspondence to intelligence. Even so, the increase in average volume from australopithecine to modern human is so great as to imply a substantial advance in intellect. Our brain case is not only large, it is distinctive in shape. Our forehead is much taller and more vertical than that of any of our predecessors. In fact, our forehead forms but one part of a face that is extremely flattened in comparison to that of an australopithecine. Our muzzle is shortened, so much so that we do not even describe it as a muzzle; in profile, it aligns roughly with our forehead above. A sharp chin is also uniquely our feature. It brings the profile of the lower jaw approximately into the plane of the forehead and front teeth. In contrast, in australopithecines and all other extinct hominids, the mouth projected forward well beyond both a receding chin and a receding forehead. The orbits, or bony rings encircling our eyes, are reduced in size compared to those of our ancestors: our ancestors were big-browed. In all species, the orbits are subjected to stress in chewing, and this stress is apparently much reduced in modern *Homo sapiens,* as can be seen also in the diminished size of our jaws. As have our jaws, our teeth have become smaller. This is particularly true of our molars, which are smaller than those of a slender australopithecine and very small indeed compared to those of a robust australopithecine.

What is known of the pelvic and limb bones of australopithecines has for some time been taken to indicate that these animals were fully erect in posture. This inference has been confirmed by Mary Leakey's spectacular discovery of tracks of fossil footprints more than three million years old. Apparently these were left by slender australopithecines treading across a fresh layer of volcanic ash in what is now Tanzania. The tracks look hauntingly like our own.

The story for upper limbs is not quite the same. Sparse fossil remains of hand bones suggest that australopithecines were not our equals in dexterity. Possibly their prominent jaws and teeth performed certain functions, such as preparation and processing of foods, that in more advanced hominids were taken over by nimble hands guided by a nimble brain.

This brief comparison sets our difficult task, which is to trace the evolutionary pathway from australopithecine to modern human. How the head and its functions have undergone their substantial changes is the primary question. To provide the raw material, we will complete our rogues' gallery of known fossil forms.

Homo erectus, as his generic name implies, was not an australopithecine but a close relative of *Homo sapiens. Homo erectus* has had many aliases. For a time he went under the generic name of *Pithecanthropus,* but this label failed to express his similarity to living humans. He traveled widely, and has been nicknamed both Peking Man and Java Man. Peking Man was the name given to a number of *Homo erectus* skulls collected after the First World War at Choukoutien, China. Tragically, these were mysteriously lost during the Second World War, though good artificial casts remain for study.

Homo erectus has a pivotal position in our story because of one simple fact. He was a distinctive species that survived for well over a million years, from more than 1.6 million years ago to at least as late as a half million years ago. Thus, he overlapped in time with australopithecines, but survived to a much more recent date. Indeed, some claim that he existed as recently as 200 thousand years ago. His earliest known occurrence is in Kenya, on the eastern shore of Lake Turkana (formerly Rudolf). He had reached Java about a million years ago, and he survived to about 400 thousand years ago in China, as what has been called Peking Man.

In *Homo erectus,* the cheek bones do not project outward as flanges, in the fashion of the australopithecines. The jaws and teeth are also relatively smaller, and the brain case is larger. All of these features place *Homo erectus* in a position intermediate between australopithecines and *Homo sapiens,* and his brain capacity also falls in the middle. Brain volume in *Homo erectus* ranged at least from 800 to 1,300 cubic centimeters, the upper limit coming well within the variation of our own species. This overlap by no means implies that some individuals of *Homo erectus* were our intellectual equals. First of all, average size is more meaningful than upper and lower limits, and here *Homo erectus* was some 300 cubic centimeters behind us. Second of all, quality counts, not simply quantity. *Homo erectus* lacked the high forehead and, we may infer, the frontal brain structure of our species. His brain case was low and long. His brow was much heavier than that of a slender australopithecine. His cheek teeth and jaws were reduced compared to those of australopithecines. The muzzle of *Homo erectus* still projected forward much father than the forehead or the point of the chin, but it may be said that this species is the earliest in which the brain case gives the appearance of dominating the skull. For the first time, the face and jaws seem subordinate; *Homo erectus* gives the impression of having

146

FIGURE 7.3

A reconstruction of *Homo erectus*, formerly known as *Pithecanthropus*. This species existed with little change for more than a million years.

been a brainy animal. What is known of his limb bones suggests an upright posture not differing greatly from our own, as would be expected from the evidence that his apparent ancestors, the slender australopithecines, resembled us in posture and locomotion.

There is a hint of an evolutionary increase in brain size within the lineage that we call *Homo erectus,* but the case is not clear. For the oldest Kenyan specimen (dated at about 1.6 million years), brain volume is estimated to have been about 850 cubic centimeters. For the Chinese skulls, which represent a relatively young population (about half a million years old), average brain volume is slightly more than 1,000 cubic centimeters. We cannot be certain that the difference here of some 20 percent represents a true evolutionary trend—we do not have enough samples to know how brain size or body size may have varied from place to place at any time. More importantly, there was no approach toward *Homo sapiens* in forehead development or, we may infer, in intellect. G. P. Rightmire has, in fact, measured several features of the skull and teeth of *Homo erectus.* He has found no evidence that any of the features, including cranial capacity, underwent statistically significant change during more than a million years of geological time. This does not rule out a small amount of change, but as far as we can tell, throughout his existence *Homo erectus* did not vary greatly in form. He was a distinctive, long-lived species. Certainly he varied somewhat in form from time to time and place to place, but less variability is displayed among all known specimens of *Homo erectus* than among the living populations of our own species.

"*Homo*" *habilis* is a more controversial species than the others I have thus far described—controversial in the sense that some workers prefer to deny its existence. They divide its alleged fossil remains, which are few in number, between the slender *Australopithecus africanus* and the more robust *Homo erectus* or assign them all to the former. In fact, we do not know that *Australopithecus africanus* or any other slender australopithecine was molded gradually into *Homo erectus* by way of what is called "*Homo*" *habilis.* In many aspects of skull shape, "*habilis*" seems strongly australopithecine in character. These aspects include a long face (the region between eyes and mouth), flangelike cheek bones, and a low forehead. I have placed the name *Homo* in quotation marks when attaching it to the name *habilis* because in my eyes, as in many others, it is fundamentally australopithecine in aspect. Even its brain size, though large, does not fall within the range traditionally attributed to *Homo.*

The specimens first assigned to "*Homo*" *habilis* were collected from Olduvai Gorge, made famous by the spectacular discoveries of the Leakey family. The specimens were found at two levels, dated at about 1.8 and 1.6 million years ago. Even if we focus just upon these remains in light of what is known about slender australopithecines and *Homo erectus*—the two groups which they have been supposed to link—we discover something quite startling about human evolution. By approximately 1.6 million years ago, when the youngest *"Homo" habilis* populations now recognized lived at Olduvai, *Homo erectus* was already in existence near Lake Turkana! Even the slender australopithecines lived on until perhaps two million years ago, and they may have cooccurred with *Homo erectus,* also near Lake Turkana, about 1.6 million years ago (the evidence is under debate). In the light of these facts, the old idea of *Australopithecus africanus* being gradually transformed into *Homo erectus* by way of *"Homo" habilis* is now difficult to defend. The slender australopithecine may conceivably have turned into *"Homo" habilis,* but the abrupt appearance and subsequent stability of the more distinctive *Homo erectus* are suggestive of punctuational transition to this younger species.

Australopithecus afarensis is the name recently suggested for some ancient, slender australopithecines. The oldest of these, which exceed three million years in age, were unearthed by Mary Leakey and her colleagues at a site in Tanzania known as Laetoli. Here the slender australopithecine apparently left the remarkable sets of footprints (tracks much like our own) that have been discovered in beds of volcanic ash. The Laetolil fossils and Ethiopian remains younger than three million years have been united under the name *Australopithecus afarensis.* Those who have proposed this name regard the populations to which they apply it as the true ancestors of the genus *Homo.* The idea is that these forms gradually turned into *Homo habilis.* It is claimed that they resemble *Homo* slightly more closely than do the slender australopithecines traditionally assigned to *Australopithecus africanus.* The latter is asserted to exhibit specializations diverging from the condition of *Homo* and its true ancestors—specializations such as strengthened jaws and slightly enlarged molars and premolars.

In truth, what has been called *Australopithecus afarensis* is a slender form that differs little from the populations known as *Australopithecus africanus.* Some would unite them as a single species. A problem with sidetracking the younger *"africanus"* populations for being slightly specialized in ways not later seen in *Homo* is that this leaves a gap of nearly a million years between *"afarensis"* and *"Homo" habilis.*

While it may nonetheless be true that *"afarensis"* and not *"africanus"* was ancestral to *Homo,* we must exercise caution in jumping to this conclusion on the basis of just a few aspects of teeth and bones. The gradualistic tradition in

FIGURE 7.4

Tracks of *Australopithecus* at Laetolil, Tanzania, in hardened volcanic ash more than three million years old.

the study of human origins for years fostered a myth that the sidetracking of *"africanus"* exemplifies. This myth is the belief that each population along the alleged continuum from ape to man must be intermediate in character between the population preceding and the one following. The punctuational scheme complicates things, but in a useful way. It allows for deviation. When one species sprouts from another, it may evolve in an altogether new direction—a direction not characteristic of earlier speciation events. Changes associated with quantum speciation reflect accidents of geographic location—the imprint of local selection pressures on a small population. They also reflect accidents of mutation—what particular chromosomal or genic change happens to provide raw material for selection within the small population. Not all paths lead toward *Homo sapiens,* and possibly no persistent path led directly toward him.

The robust australopithecines represented an unusual offshoot. If their origin was punctuational, it exemplifies what I am describing. The Neanderthals, which I will discuss next, display strange specializations that even now cannot be explained, let alone projected from what went before, and these specializations do not appear in modern human populations. Whether Neanderthals were our ancestors or not, they were in some ways anomalous in the general context of hominid evolution, and I would regard our own populations as being in some ways even more aberrant, but in quite other directions. Out of nowhere, our sharp chin, weak brow, and high vaulted forehead appear in the fossil record. These particular features are utterly unpredictable on the basis of what preceded them. Finally, there is a particularly telling example of inconsistency in the direction of human evolution. The brow of *Homo sapiens* is less robust than that of *Homo erectus,* yet *Homo erectus* is more heavily browed than its slender australopithecine ancestors. This slender-to-robust-to-slender pattern is a clear example of evolutionary reversal, whatever the details turn out to be. Viewed in this light, the minor deviant specializations of *Australopithecus africanus* should not automatically disqualify this temporally intermediate species from an evolutionary position between the populations labeled *"afarensis"* and the much younger genus *Homo.*

Whatever may have been the ancestral role of *Australopithecus afarensis,* this species offers further justification for favoring the punctuational model. We can note that even as originally defined, on the basis of material from only two sites, *Australopithecus afarensis* existed for several hundred thousand years with only modest change. This species has been characterized as having been very primitive. Whatever species assignments are favored, slender australopithecines in general persisted from considerably more than three million years ago to two (or possibly even 1.5) million years ago, without substantial restructuring. If we could project these rates through about twice as much time—right up to the present—we would have a slender australopithecine walking around today! Somewhere, somehow, evolution has done much more.

Homo neanderthalensis is the formal name that I will employ for the familiar Neanderthals, whose short stay on earth lasted from perhaps 100 thousand to 35 thousand years ago. This species designation runs counter to the prevailing opinion, but the prevailing opinion seems to me to reflect strong biases. These relate to gradualism, the single species hypothesis, and anthropocentrism.

Proponents of the single species hypothesis have argued that Neanderthals had to belong to *Homo sapiens,* as a subspecies or variety, simply because they overlapped in time with *Homo sapiens.* With the demise of the single species hypothesis, this argument fell apart. Neanderthals had large brains—slightly larger than ours. Can we admit to the existence of a separate species of humans with larger brains than ours? Better that we include them with our species so as not to jeopardize our intellectual preeminence: thus runs another line of thought that has colored some thinking. As an example, one authority, perhaps under the influence of gradualism and the single species hypothesis as well, recently wrote tersely of Neanderthals, "They must be called *Homo sapiens* on the basis of brain development."

Finally—and I am inclined here to ask whether sexual fantasies have been at work—there is the notion that *Homo sapiens* of our type must have interbred with Neanderthals. Given the great breadth of human sexual behavior, interbreeding probably did take place, but we cannot be sure that resultant offspring were healthy, fertile, or socially acceptable even to their parents.

An overemphasis on brain size has obscured the fact that Neanderthal skeletons are quite distinct from those of *Homo sapiens.* A rank amateur can see conspicuous differences at a glance. These differences are great enough that comparable ones in any other family of mammals would be seen as requiring separation at the species level. A lion and a tiger have more similar skeletons, for example. I, therefore, suggest that we cast off the biases that have led us to embrace Neanderthal as one of our own and instead give him the recognition he was commonly accorded in past decades: status as his own species, *Homo neanderthalensis.*

Neanderthals, as most people know, were stocky creatures compared to us. They were not, however, stooped at the shoulder and bent at the knee, as sometimes reconstructed. Part of the problem here is that the La Chapelle-aux-Saints Neanderthal, a middle-aged male whose skeleton attracted much attention early in this century, misled his chief interpreter. As it turns out, this particular Neanderthal was slightly deformed by arthritis! The limb bones of Neanderthals were heavier than ours, and it is inferred that their grip was more powerful. Apparently, they were in general considerably stronger for their height. Both males and females had pelvises somewhat different in form from ours, and the Neanderthal shoulder blade was also distinctive. In comparison to *Homo sapiens,* Neanderthals were built for strength rather than swiftness. We must be much better long distance runners.

The most familiar characteristics of Neanderthals are the attributes of their skulls. The well-known heavy brow, like the robust body, was inherited from *Homo erectus.* The long, low, sloping skull, which lacks the high-vaulted forehead of our species, is also reminiscent of *Homo erectus,* but the braincase is larger. In fact, average brain volume for some Neanderthal populations seems to have been greater than in our species. This does not mean that Neanderthals were smarter. The difference in volume is slight and may relate to the need for extra neurons to control the Neanderthals' more massive musculature. Also, as I will describe shortly, the uniquely well-developed frontal region of our skull may relate to our intellect.

There are other distinctive features of the Neanderthal skull. The chin is variable in its degree of development, but is on the average much less sharp than ours. The cheek bones are not as high as in our species, but slope backward, and the facial region between mouth and eyes is quite large. The teeth are positioned forward in Neanderthal, leaving a conspicuous gap in the rear of the jaw. The base of the skull also differs from the base of ours in ways reminiscent of *Homo erectus.*

Somewhere, probably not much earlier than 100 thousand years ago and possibly much later, *Homo sapiens* came into being. The oldest known fossil remains of truly modern humans are about 40 thousand years old. Thus *Homo sapiens* may have overlapped with the youngest Neanderthals, although some researchers still believe that the entire Neanderthal lineage was transformed into the modern human species. How such a transformation would have taken place is difficult to imagine. In Europe, where Neanderthals had been flourishing, they were suddenly replaced by humans of the modern type between about 40 thousand and 35 thousand years ago. An evolutionary transition would be tantamount to the biological rebirth of a species. Until 35 thousand years ago, *Homo neanderthalensis* had existed for perhaps 65,000 years with no visible change. From 35 thousand years ago to the present, *Homo sapiens,* in turn, has persisted with no apparent change. If our species evolved about 35 thousand years ago, it could only have emerged by a quantum speciation event from a small population of Neanderthals. In no reasonable way can we envision a transformation of the entire large Neanderthal lineage so abruptly. The plausible alternative is that *Homo sapiens* evolved from a single Neanderthal population or from some other lineage descended from or even representing *Homo erectus.*

In any event, large populations of Neanderthals suddenly disappeared from Eastern Europe about forty thousand years ago, and from Western Europe about five thousand years later, and humans of modern form appeared in great numbers. This circumstantial evidence suggests that the transition from Neanderthals to modern humans in Europe may have involved active displacement—internecine war, a battle of the species, which proceeded across Europe from east to west. Here we may have a single instance in which the single species hypothesis approximates the truth. *Homo sapiens* was seemingly incompatible with, and perhaps militarily superior to, the older, big-browed species. Soon after coming into being, our species had the world to itself.

Looking back over the various members of the human family just described, we see that the old connect-the-dot approach to human evolution simply will no longer work. The dots were bits and pieces of humanoid remains that, in blissful ignorance, we could once align according to our gradualistic preconceptions. The dots have now become lines of descent that in many cases depict little change over long stretches of time and that sometimes even overlap. The more fossil evidence we accumulate, the longer become the durations of the little-changing entities that we call species. Thus, *Australopithecus afarensis* is considered to have survived for several hundred thousand years, but even this great duration is based on the evidence of only two collecting sites. Presumably some older and younger populations, if known, would also be assigned to this chronospecies. Our information for *Homo erectus* has particularly strong punctuational implications because this species is relatively young and belongs to our own genus. We must now contend with the temporal overlap of *Homo erectus* and australopithecines and with the stability of this species for perhaps a million years.

As it turns out, hominid species whose longevities have not been terminated by extinction have existed longer, without evolving enough to deserve new names, than have species in many other families of mammals. In other words, while the human family is supposed to have evolved with extraordinary rapidity, the fact is that well-established human species have evolved relatively slowly—or at least not any more rapidly than the average rate for other mammals!

If we shift to a punctuational interpretation of our ancestry, then we can formalize another point made earlier. We can expose what we might call "the fallacy of the missing link," or the mistaken idea that two species that existed at different times and did not interbreed can be connected only by species of intermediate form. In fact, as we have seen, reversals of evolution are not only possible, they are clearly indicated, as in the sequence: weak-browed australopithecine⟶heavy-browed *Homo erectus*⟶very weak-browed *Homo sapiens*. We cannot expect to find evolutionary reversals that precisely duplicate what went before, but we can expect partial reversions, as in the preceding example or in cases involving simple changes in tooth proportions. Even within a small portion of the tree of life, evolution does not move inexorably in one direction.

Inasmuch as the human family has undergone much total evolution, we are entitled to ask exactly what changes may have taken place by punctuational steps. We obtain some indications simply by contrasting certain species with their apparent ancestors. Early in the history of the Hominidae, or slightly before the family evolved, the origin of upright posture was a critical evolutionary step. As far as we know, a fully two-legged stance preceded the advent of sophisticated tool-making. Owen Lovejoy has suggested that this posture served to free the hands for another purpose: it permitted a mother to care for an infant while simultaneously foraging for food. I shall comment further on this topic shortly. The organs of speech may also have evolved in steps. Almost certainly the development of speech at some point in our ancestry was one trigger for enlargement of the brain, opening new directions for mental devel-

---

FIGURE 7.5
Humanlike head of a fetal orangutan, figured by Darwin.

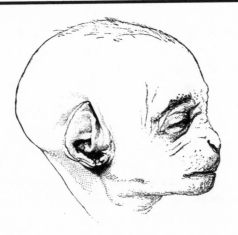

opment. It has been argued that the Neanderthals' throat structure precluded sophisticated speech, but the relevant anatomy of Neanderthals has not been sufficiently reconstructed to justify this conclusion.

Many fundamentally human features have resulted from the evolutionary pattern known as neoteny, or evolutionary retardation of the development of bodily features. We resemble a baby chimpanzee more than we resemble its mother. Like a newborn gorilla, we are largely hairless except for our heads. Our hands and feet bear pudgy digits and soft pads; they look much like the extremities of a fetal ape. These and a number of other human traits indicate that an important aspect of our evolution has been a kind of juvenilization, whereby embryonic or youthful traits of our ancestors have been shifted to the adult stage of our development. As a result, we are born at an unusually immature stage: a human baby in many ways has the form of an embryonic monkey. Apparently one of the adaptive reasons for our neotenous evolution is that it has served to enlarge our brain—baby hominids have larger heads for their bodies than do adults. The helplessness of the immature human infant has important implications. We must care for our offspring for a long period before they can fend for themselves. At the same time, of course, we train our offspring, and this is a matter of passing on the elaborate cultural tradition that our big brain has engendered.

I venture to suggest that neotenous enlargement of the brain somewhere in the ancestry of the australopithecines was closely linked to the evolution of upright posture. Monkeys and apes are sufficiently hairy and their offspring sufficiently mature at birth, that an infant can successfully cling to an ambulatory mother. Neoteny has rendered this kind of infant behavior impossible for modern humans. The problem is two-fold. Our mothers are relatively hairless and our infants are too immature to cling. Of necessity, the cradling arm has taken over (and with it, the pendulous breast). One possibility is that upright posture preceded and made possible the juvenilization that has been so important in our evolutionary history. Just as likely is the possibility that two-legged posture was not well developed before neoteny had proceeded very far. If this was the case, neotenous evolution must have dictated the evolution of upright posture. Infants as helpless as ours could not cling to a quadrupedal mother, especially one lacking both fur and the dexterity to fashion clothing.

Clearly some neotenous change occurred before the rise of *Homo sapiens,* but it seems evident that our species may itself have arisen by an event of quantum speciation in which neoteny played an important role. It has long been recognized that our flattened face and lack of brow ridges are neotenous traits. In fact, the skull of the young Neanderthal and the young *Homo erectus,* like the skull of the young ape, lack heavy brow ridges. Our unique, high-vaulted forehead, which forms part of our flattened face and may be related to our brainpower, is mimicked in embryonic apes.

The question that we must ask is whether the neotenous step in the origin of our species was guided primarily by natural selection for a large, frontally expanded brain. Is it possible that, as a part of juvenilization, other skull features—ones that happen to be genetically linked to frontal brain development—were simply dragged along? Are all of the neotenous features of our head and body advantageous? At present, we simply do not know.

We also do not know to what degree the general increase in brain size that preceded the frontal expansion of *Homo sapiens* may have occurred by gradual evolution within established species.

Homo sapiens exhibit an extraordinary range of variation in form for a species of mammals. Much of the variation is, of course, expressed as differences among races. The fact that our races have tended to remain distinct until recently is an indication that advanced hominids have tended to remain subdivided: gene flow has probably not been able to spread useful new genetic features efficiently throughout most species. This is one of the factors that I have invoked to explain the slow transformation of species in general. In other words, we would not expect a species like ours, with its varied languages and cultures, to evolve effectively as a unit, and the brief history of *Homo sapiens* bears out this prediction.

The great flexibility of hominid sexual behavior may have tended to promote some interbreeding between races and even between species, but cultural contrasts have presumably in many cases tended to prevent the complete fusion of races. In fact, the unique complexities of human communication and culture have probably promoted punctuational steps by creating incompatibilities between breeding groups. Many punctuational steps in human evolution may have represented subspecific degrees of change—the origin of distinctive races. Only occasionally has there occurred a truly dramatic step, such as I would invoke for the origins of *Homo* and, later, of *Homo sapiens*.

How might *Homo sapiens* have come into being? Here we can engage in a bit of plausible fantasy—fantasy constrained by evolutionary theory. Envision within a robust species of the genus *Homo* a small group of individuals—perhaps a handful of close relatives—possessed of physical, behavioral, and intellectual traits that stand out as bizarre among their tribesmen. These unusual individuals lack the handsome, heavy brow ridge that adorns normal individuals of their species. Instead, members of the small group are dish-faced and have unbecomingly tall foreheads. The traits of the individuals result from a small number of mutations that have induced a slight neotenous transformation. We might suppose that, as a result, the transformed individuals are also curiously less hairy than their tribesmen. Possibly the pariahs are banished from the tribe as ones possessed by demons. Possibly they steal off at the dark of the moon to escape persecution. It happens, however, that what their former compatriots have seen as freakish traits are actually, for the time, marks of genius. Having no alternative in their isolation and perhaps finding mutual attraction in their similarity, the exiled few embark on a new tribal life, possibly (but not necessarily) in a new kind of habitat that demands new modes of food gathering or defense. They retain the sex drive. They breed. Their children and their children's children thrive, owing in part to the inheritance of an anomalously large frontal region of the brain and extraordinary intellect. Inbreeding leads to the expression of certain recessive genes that were present but unexpressed in the small founder population, and some of these turn out to be advantageous and are fixed. In the course of a few generations, or perhaps a few hundred, the modified descendants of the unusual founder population represent a new species. In time, it comes to call itself *Homo sapiens,* the wise human.

We do not, in fact, know that we represent a more intelligent species than Neanderthal. The frontal lobes of our brain are expanded, but the functions of these are poorly known. While we have no proof, we probably are not misguided in assuming intellectual superiority for ourselves. From impressions of the brain on the skull, it has been shown that the so-called basal neocortex, which lies at the foundation of our frontal lobes, is uniquely well developed. The skull configuration of *Homo erectus* and of Neanderthal reveal that this part of their brain was less well developed. This disparity is of no mean significance. The basal neocortex has been shown to govern our ethical judgment and individual behavior in social relationships. In short, it determines most of what we call our "personality." Certainly the emergence of modern culture may relate to enlargement of the basal neocortex.

Whatever the exact story of our ancestry may be, the idea that our species may have emerged from an exiled social group has one startling implication. The founder population may have created its isolation by choice—by a reasoned decision. It lacked the vaguest knowledge of organic evolution and was by no means charting its evolutionary course, but it may unwittingly have made the choice that led to its rapid transition into the human species.

Members of the human family have evolved culturally, of course, just as they have evolved biologically. Cultural traits, such as methods of producing implements or obtaining food, are inherited, but with much greater flexibility than are biological traits. Despite obvious differences between the two kinds of evolution, it is interesting to seek out parallels and analogies. In particular, we can inquire whether cultural change may to any degree have followed a punctuational pattern and, if it has, whether periods of stagnation have tended to coincide for the two modes of human evolution. On the other side of the coin, have rapid biological and cultural transformations gone hand in hand?

Thirty-five thousand years ago, when Neanderthals disappeared, *Homo sapiens* suddenly became widespread in Europe in the form of the Cro-Magnon people, who were anatomically indistinguishable from modern Europeans. Since that time, *Homo sapiens* has obviously undergone extraordinary cultural change while remaining virtually stagnant in anatomical configuration. Thus, in our own species, we find no indication of correspondence between biological and cultural change. Earlier in human evolution, however, the picture is very different.

As we have seen, *Homo erectus* was a long-ranging biological species, spanning more than a million years and disappearing no earlier than perhaps 500 thousand years ago. This robust species employed fire—though for what variety of purposes we do not know—and his butchering sites have been found in several widely separated areas. Horses, deer, rhinoceroses, and elephants were apparently among his prey. Their bones are associated with his stone hearths. In Spain, there is evidence from two sites that *Homo erectus* trapped animals in bogs and butchered them there. He apparently fashioned implements of polished wood and bone, and his characteristic stone artifacts are labeled Acheulian, after the locality in France where they were first discovered. Acheulian artifacts are bifacial, which is to say they were formed by chipping opposite sides of a stone to produce a sharp edge. What is remarkable is that the manufacture of Acheulian tools spanned a long interval, from about 1.4 million to 200 thousand years ago. Like the anatomy of *Homo erectus,* the Acheulian industries show only modest change from place to place and from time to time. Because *Homo erectus* did not bequeath us burial sites linking him to particular artifacts, we do not know that he produced many tools labeled Acheulian. It is interesting, however, that the beginnings of the Acheulian industry studied at Olduvai Gorge by Mary Leakey are estimated to be slightly older than 1.6 million years, or are approximately contemporaneous with the oldest known skulls of *Homo erectus.* These beginnings are recognized within what is labeled Developed Oldovan culture. They are seen in the form of primitive bifaces—imperfectly crafted and scarce examples. We do not know whether an earlier species initiated ths Acheulian industry, but it is quite possi-

FIGURE 7.6
Acheulian hand axes produced by *Homo erectus*.

ble that fully developed Acheulian culture was largely linked to the biological entity we call *Homo erectus*. Both persisted for most of the Ice Age.

Looking backward, we find that the oldest apparent stone artifacts now recognized are pebbles that were chipped crudely into what may have been cutters and scrapers. They were not weapons, as many bifaces may have been. Their age is debated, but certainly approaches two million years. Part of the impetus for placing the species "*habilis*" in the genus *Homo* has come from the temporal overlap between this problematical form and stone tool cultures at Olduvai, but this linkage is unjustified. We flatter ourselves unduly with the notion that a manufacturer of tools automatically belongs to our genus. Australopithecines may have ushered in the Stone Age.

Looking beyond *Homo erectus* toward the present, we find that the Neanderthals also had a characteristic culture, the Mousterian. Mousterian industries produced flakes of flint for projectile heads, scrapers, and knives. Neanderthal's culture, like his anatomy, was notably stable throughout his existence.

By about thirty-five thousand years ago, the Mousterian had given way in Europe to the more advanced Upper Paleolithic cultures associated with skeletons of modern *Homo sapiens*. As I have noted, the transition apparently swept from east to west in Europe between about forty thousand and thirty-five thousand years ago. Early Upper Paleolithic artifacts include elements of Mousterian culture, indicating that Neanderthals passed at least some of their culture to modern humans. It is, of course, uncertain whether this inheritance was

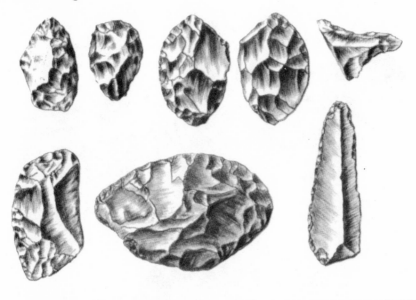

FIGURE 7.7
A variety of stone implements fabricated by Neanderthal.

associated with biological transition or was a matter of secondary contact of the two species and cultural assimilation by *Homo sapiens* while it displaced the Neanderthals.

Within the human family, cultural traditions must have altered more easily than important features of the genetic code. Although human culture must generally have been more labile than human biology, the mere fact that we see evidence of cultural stability in groups like *Homo erectus* and the Neanderthals raises questions of analogy. The heavy brow and low, sloping frontal regions of the crania of *Homo erectus*, Neanderthal, and all other known forms older than about forty thousand years seem to reflect certain inherent limitations of the enclosed brain. In one and one-half million years, these forms never progressed beyond a primitive Stone Age culture. Sophisticated art seems to have formed no part of their life. Cave painting and other advanced forms of decorative representation appeared only with the rapid cultural advancement of *Homo sapiens*. Is it possible that the evolutionary step that gave us the high-vaulted forehead of the juveniles of our ancestors also transmitted to our adulthood the youthful traits of inquisitiveness and imagination?

Even if *Homo erectus* was intellectually incapable of moving beyond a primi-

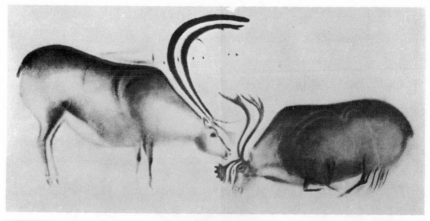

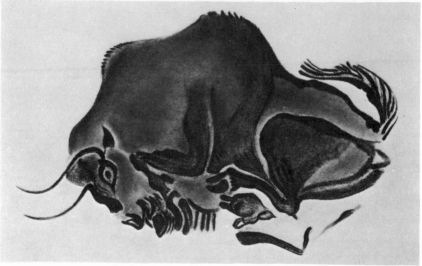

FIGURE 7.8

Art of early Cro-Magnon (early *Homo sapiens*). Above: Male and female reindeer, Fonte de Gaume, France. Below: Wisent (European bison), Altamira Cave, Spain.

tive Stone Age level of culture, we may ask why, at this level, he did not display greater flexibility. Perhaps before the creative brainpower of *Homo sapiens* was unleashed, cultural evolution within established species of hominids was also subject to braking mechanisms analogous to those that tend to stymie biological evolution in large species. Complete transformation of an entire culture, like the anatomical restructuring of an entire species, must somehow involve many populations. A dramatically new feature must either crop up everywhere, or it must spread from tribe to tribe—in the biological case by sexual, and in the cultural case by social, intercourse. But mixing of genes or ideas is not enough. Cultural innovations, like biological innovations, will only supplant what preceded them if they are of value. In the case of culture, this is a matter of judgment and conscious decision. What then might militate against cultural change? Mores, religious beliefs, parentally emplaced psychological strictures, and rational fear of the uncertain are comparable to the constraints on genetic change within a population. Most accidental changes in an established culture, like most genetic mutations, must be deleterious. This we are told both by our emotions and by our reason. Cultural systems that we have found to work can only be altered at some risk. In this light, is it really surprising that our less innovative ancestors bore the stamp of cultural conservatism? Certainly both emotions and reason may have reined in whatever creative drives stirred within these early bearers of culture.

Only in our species has cultural change accelerated at an alarming pace. Obviously, level of culture has been tied to level of intelligence in a general way within the Hominidae. On a finer scale, and especially with our own culturally diverse species, this generalization breaks down. Alfred Russell Wallace, whose independent conception of natural selection forced Darwin into print in 1859, took the opposite point of view. Wallace was a strict gradualist who assumed that every basic human feature must be under the tight control of natural selection. He observed that groups of modern humans resemble each other in level of intelligence, yet differ greatly in level of cultural development. His conclusion was that natural selection, although it produced all lower forms of life, could not have developed the human mind. If it had, Wallace reasoned, a high level of culture should characterize all populations of our species; as things stand, however, intelligence is being wasted in modern Stone Age societies. Where Wallace erred, of course, was in assuming that culture is tightly linked to biologically inherited traits. In reality, a single Stone Age person may be using mental powers equal to those employed by a nuclear physicist, but in a different context. Was the inventor of the wheel not thinking just as hard, when there was no wheel, as the physicist who adds a small increment of knowledge to an already sophisticated body of science?

The fact that biological and cultural evolution are not tightly intertwined underscores the point that the application of punctuational schemes to cultural evolution is largely a matter of analogy. The fact that some species have been characterized by particular cultures relates to the fact that a social group with a unique culture is also an interbreeding group with its own gene pool. Thus, it is to be expected that some speciation events will approximately coincide with cultural transformation.

Looking backward, it seems clear that our species' study of its own evolution has not been without bias. As I have noted, it was partly an intellectual climate of gradualism and partly a sparseness of relevant fossil data that until recently prevented even the biological evolution of humans from being viewed in a punctuational context. There is also something special to us about our own evolution that seems to have fueled the gradualistic fire. The single species hypothesis is in a sense self-congratulatory. It sustains the view that we are the pinnacle of evolution—the ultimate product of a unique and superior lineage.

In the *Origin of Species,* Darwin carefully avoided the inflammatory subject of human ancestry. Nonetheless, his publication led immediately to a widespread public preoccupation with our alleged simian heritage. One Victorian lady is reputed to have responded: "Descended from the apes! My dear, we hope it is not true. But if it is, let us pray that it may not become generally known." Soon, however, T. H. Huxley showed that, anatomically, humans resemble apes more closely than apes resemble monkeys, and there emerged a formidable burden of evidence that apelike animals were indeed our recent ancestors. The strong gradualistic belief that then developed—the belief that a single lineage led without discontinuity or reversal to a single modern human species—offered some accommodation to the uncomfortable idea of bestial ancestry. It represented a way of salvaging some dignity. If we no longer stood at the top of a fixed *Scala Naturae,* closest of all earthly beings to God, we at least represented the culmination of the perfecting process of natural selection within a very special line of descent.

The science of anthropology was slow to take root. It emerged as a distinct discipline only late in the nineteenth century. When the *Origin of Species* was published, the only hominid fossil known was a Neanderthal skull discovered in Germany in 1856, and this solitary fossil could easily be dismissed as a victim of the biblical flood or a diseased or freakish human. *Homo erectus,* which at first went under a different name, was unearthed in Java during the 1890s, but all other hominid finds have come in the present century. Physical anthropology, or human paleontology, is a science heavily dependent upon chance discovery. The details of our ancestry remain uncertain, but we know far more about our family tree today than was apparent two decades ago, and as discoveries have extended the recognized time spans of fossil species, our family tree has taken on an increasingly punctuational shape.

8

Continuity or Creation?

T HERE IS no doubt that the new punctuational movement will bring joy to the hearts of creationists—those who claim species to be discrete entities that a divine being brought separately to life and placed upon the earth. The fossil record, in offering the punctuational message that distinctive forms somehow appear suddenly and, once established, change slowly, would appear to be playing into the creationists' hands. It might then seem an unfortunate quirk of fate for evolutionists that the punctuational movement coincides with a religiously-motivated drive to revive creationism as a legitimate field of science and to force our public schools to teach the particular story of creation favored by fundamentalist Protestants.

Viewed in a different way, however, the coincidence may actually be seen as fortunate: if creationism had to rise again, it is well that we have punctuationalism to counter some of its arguments. Many of the complaints that creationists have leveled at gradualistic evolution (complaints that this model of change does not square with the facts of the fossil record) now appear baseless. With the acceptance of the punctuational scheme, the sudden appearance in the fossil record of many distinctive groups of animals and plants need trouble evolutionists no longer.

A lingering question is whether the punctuational view nonetheless opposes the ideas of the Modern Synthesis forcefully enough to throw evolutionists into such disarray that the very idea of evolution must lose credence. Despite what creationists might claim, such an attitude would amount to throwing the baby out with the bathwater. It is true that biologists and paleobiologists debate the patterns and the mechanisms of evolution, but they nevertheless stand overwhelmingly in favor of the idea that evolution has occurred.

The punctuational view does not, in fact, push us to the brink of creationism. The question here is one of continuity of descent. Certainly, if evolution moves with a pulsating tempo—if most change is confined to small, rapidly diverging populations—then we are forced to confront the issue of continuity. Are there complete breaks in the flow of generation-by-generation evolutionary change? Do distinctive new species appear full blown in the form of single individuals or pairs of individuals bearing suites of new organs and behaviorial patterns that, even at the outset, are well adjusted to the environment? Is there comparatively little change in the generations that follow?

Our answer is that distinctive new species are not literally born or hatched in final form. It is virtually inconceivable that the first bird emerged full blown, from a dinosaur egg, or that the modern giant panda entered the world as a monstrous bear cub. Certainly, however, a partial step in such a direction can be taken in a single generation. As we have seen, a small group of siblings may, for example, share certain features that set the stage for rapid divergence. For speciation to be achieved, however, it is required that such features be fixed within an interbreeding population and that they be blended with other adaptations to yield a successfully functioning unit of life. This may require several generations—or several hundred or several thousand. Such intervals are nonetheless brief instants in geological time, and this is the fundamental point of the punctuational model of evolution. Once established, an average species survives, as a slowly evolving lineage, for at least a million years (as in mammals, birds, and insects) and for more than ten million years in some groups (snails, clams, and corals, for example).

Quantum speciation entails no major elements not recognized within the Modern Synthesis of evolution. The new view simply differs in its emphasis on particular elements and, as I will show in the following chapter, in its implications for large-scale evolution. Accidents of genetic composition and of breeding play a greater role in quantum speciation than in the evolution of large populations. Selection is important in quantum speciation—but a coarse-grained kind of selection operating within very small populations. Here unusual new features loom large. The small size of the population has two major effects: breeding often serves to spread the new traits throughout all members of the population, and chance is also accorded a larger role.

In Darwin's day, religious convictions were the primary source of opposition to the idea of organic evolution. The same is true in the modern world. Today, staunch creationists are rare outside the United States. They are also rare beyond the bounds of fundamentalist Protestantism. These simple facts are sufficient to condemn the present creationist movement as unscientific. Objectivity is essential to science. Religion is not science, and if a particular set of religious views is found to be the sole animus behind an idea, then science cannot foster that idea—though it can actually reject the idea only if scientific procedures refute it. As I will point out, scientific procedures do disprove the central tenets of the modern biblical creationist movement. Genesis cannot be literally true.

Evolutionists, unlike fundamentalist creationists, represent a wide variety of religious viewpoints. Some evolutionists are active Christians; some practice other religions, and others are agnostic or atheistic in outlook.

Interestingly, it was Darwin's champion, T. H. Huxley, who invented the word "agnostic," meaning "having no knowledge" (or pleading ignorance on the subject of a divinity). Agnosticism is, in fact, the only position that is tenable *within* science, and Huxley was strictly a scientist. Atheism, or denial of the existence of a divinity, is as illegitimate within science as is theism. Science simply does not deal with divinities. Observations and measurements are the raw material of science. Revelations, both personal and spiritual, and faith are the raw materials of religion. In each area, logic shapes the raw materials into general beliefs, but separate foundations lead to mutually exclusive domains. All of this means that scientists may be religious—and many evolutionists are—and yet their religious views, having a separate basis, cannot legitimately impinge upon their science.

As I have noted, Darwin set out on the voyage of the *Beagle* as an orthodox member of the Church of England with the goal of documenting the literal truth of the Bible. He was also a firm believer in empirical science, and when his observations conflicted with the idea of Divine Creation, he reluctantly abandoned his Christian orthodoxy. His belief in a divinity eroded more slowly. There is some evidence that Darwin's theism remained in some form even when he penned the *Origin*. Later in life he became an agnostic, though retaining suspicions about the need for a prime mover in the history of the universe.

Interestingly, the Roman Catholic Church, unbound by literal interpretation of scripture, does not oppose the concept of evolution. God created man, is its position, and to most interpreters this has implied the injection of a soul into a seminal population of our species.

FIGURE 8.1

A reconstruction of the ritualistic burial of the young Neanderthal discovered at Le Moustier, France.

Continuity or Creation?

Science does not traffic in souls, but it does treat questions of cultural attitudes toward them. I cannot help but pose a question for those who might see *Homo sapiens* as uniquely possessing a soul. I have argued that Neanderthal was a distinct species, not a member of *Homo sapiens*. I have no idea whether Neanderthal had a soul or not, but he thought that he did! This, anthropology tells us with a high degree of certainty.

In 1908, at Le Moustier, France, the town that gave Neanderthal culture its name, the skeleton of a teenage Neanderthal boy was discovered reclining peacefully on one side, with his right arm crooked beneath his head, in what was obviously a shallow grave. He was buried with flint chips surrounding his head and stone tools flanking his body. Also interred with the skeleton were charred animal bones that apparently once bore a supply of cooked meat. In a cave near the same town, the skeleton of a Neanderthal infant was unearthed from one of nine mounds arrayed in three staggered rows. Accompanying the infant were three flint tools. In the Zagros Mountains of Iraq, a Neanderthal man who had died of a crushed skull was found buried tenderly on a bed of boughs and flowers, many of which can be identified today from the resistant pollen grains that they left behind.

These burial sites reveal that some members of the Neanderthal species revered their dead and almost certainly expected that they would have an afterlife. In other words, it may be that there have been two species of the animal kingdom who have had among their members individuals who have thought themselves immortal in spirit. We must ask ourselves, do our religions have a right to claim uniqueness here? Can any species deny another species a soul with any more certainty than the other species can lay claim to one? I do not know what any modern religious organization would see fit to do with the idea that Neanderthal, with his implicit beliefs, was a discrete species, but the evidence that suggests that he was exactly that gives us cause for reflection.

Creationists' opposition to evolution raises the question of whether evolution is a fact—whether it has been proven. Most evolutionists would argue that it is almost certainly a real phenomenon, and a phenomenon powerful enough to be responsible for the varied forms of life we see around us. Absolute proof is another matter. Many of us adhere to the idea that science never proves anything. It provides no more than a very high degree of certainty. The connection between cigarette smoking and cancer offers a familiar example.

An enormous body of circumstantial evidence points to smoking as a cause of lung cancer. Statistical treatment of the incidence of lung cancer in smokers and in nonsmokers shows that there is only the slimmest of chances that smoking is not linked to cancer. Statistical treatments never offer proof, however; they simply give estimates of probability. So high is the probability in the smoking example that it would be hard to find an unbiased scientist who, after viewing the available data, would not bet on the presence of a connection.

The inherent lack of absolutes in statistical analysis offers the tobacco companies an escape, however. They can claim that no causal connection is proven. They are on safe ground on two counts. Only the first has to do with the impossibility of statistical proof. Here they simply avoid telling us that the probability of no connection is a tiny fraction of 1 percent. The tobacco industry's second escape route has to do with causality. Even strong evidence of connection does not establish an explanation. Perhaps the connected traits are not directly related as cause and effect. Perhaps people who tend to smoke share certain personality traits that happen also to confer a special vulnerability to cancer. This, of course, *seems* unlikely, but we cannot rule it out altogether until we have good reason to believe that it is false. What truly constitutes demonstration of causality? If it were found that smoke caused certain cells in the lungs to change their appearance, and if these altered cells were observed usually to become cancerous, the tobacco companies could still argue that there was no real proof of a causal relationship. The case against smoking would still be statistical. The next demand might be for an apparent biochemical explanation. Even if this were forthcoming, it would not constitute proof, but only a likely mechanism. In short, if they choose to hide behind the requirement for absolute certainty, tobacco companies will always be able to make their present claim that, despite what the Surgeon General may say, smoking has not been proven harmful to your health.

Many students of biological systems are even more certain that evolution has occurred than that smoking causes cancer, but because they believe that science does not prove theories, they are barred from claiming absolute proof. The classic example of our inability to prove in science relates to the process we call induction—the extraction of general principles or theories from bodies of data. A common cliché here is that we do not know with absolute certainty that the sun will appear in the East tomorrow. Throughout recorded history, the sun has shown in the East every morning. This means that it almost certainly will make an appearance tomorrow, but we have no proof: we cannot generalize that the sun will *always* appear in the East. Still, who would wager that the sun will not rise tomorrow?

Recognizing that absolute proof is not a legitimate issue, we then ask ourselves how many biologists untouched by religious fundamentalism do not consider evolution a near certainty. The answer, of course, is "very few." How has this verdict been reached?

Although science does not prove, it does disprove. When a theory with many implications has withstood the threat of disproof for many years, it is granted a very high probability of being valid: it gains general acceptance, if not proof. There are two ways that a theory can be refuted. One is by the discovery of direct evidence opposing it. The second is by refutation of its corollaries or predictions. What I mean by corollaries or predictions is logical implications, which can be independently tested. When a scientific idea has a large number of implications, it is said to have high information content. If we test many of these over a period of years and find no contradictions, the basic idea becomes enriched and strengthened by its historical success. This is exactly what has happened to the general idea of organic evolution. For more than a century, it has offered an enormous variety of testable predictions, yet none of these has been called into question to the degree that evolution has lost general support. On the contary, evolution is more popular among scientists than it was in the decade after Darwin published *On the Origin of Species*. Details of the pattern and mechanism of evolution have been debated, but the general idea has, in the popular phrase of science, stood the test of time. As we have seen, it suffered some defections in the first decade of the twentieth century, but then rebounded to a stronger position than it ever held before.

There is an infinite variety of ways in which, since 1859, the general concept of evolution might have been demolished. Consider the fossil record—a little-known resource in Darwin's day. The unequivocal discovery of a fossil population of horses in Precambrian rocks would disprove evolution. More generally, any topsy-turvy sequence of fossils would force us to rethink our theory, yet not a single one has come to light. As Darwin recognized, a single geographic inconsistency would have nearly the same power of destruction. An isolated natural occurrence in California of several species of kangaroos identical to those of Australia would create havoc with our thinking, but no contradiction of this magnitude is known. Several less dramatic disjunct patterns of distribution that once seemed somewhat puzzling have been resolved by independent evidence that continents have moved in ways that account for the facts. Even a successful scientific idea nearly always follows a zigzag upward course in running the gauntlet of time.

The general concept of evolution has not merely resisted refutation, it has gathered strength from new developments. As we have seen, fossil evidence that once seemed to indicate the almost instantaneous appearance of diverse groups (of the earliest marine life of the Cambrian, for example) has given way to more detailed fossil information that documents intervals of diversification. Geographic patterns, which as much as anything accounted for Darwin's conversion to evolution, have, on closer inspection, become even more striking. We now recognize populations that form rings around inhospitable territory and that show stepwise geographic change of form that clearly has resulted from evolutionary divergence as populations of the initial species have spread clockwise or counterclockwise around the barrier. When the barrier has been fully encircled, it has often turned out that the youngest populations are distinct from the oldest populations, with which they now make contact. Progress around the barrier has obviously been marked by evolutionary steps. The historical enrichment of the general theory of evolution is also evident in the growth of modern genetics, which swept aside the temporary obstacle of blending inheritance and went on to offer new levels of evolutionary understanding. In the last century, who except Gregor Mendel ever anticipated the discovery of particulate inheritance and its great potential for explaining the variability that constitutes the raw material of evolution? It has now actually been shown that this variability can be put to evolutionary use. For example, new species of fruit flies—populations that cannot breed successfully with their ancestors—have been created in the laboratory by stringent artificial selection.

The fundamentalists who still argue for "scientific" creationism are not arguing for anything scientific at all. Their approach is to attack the foundations of evolutionary theory, and their claim is that when evolutionary theory crumbles, creation will somehow stand confirmed in its place—not creation in general, but the particular account of creation that appears in the Bible.

Divine Creation is not a single concept. It represents a spectrum of possibilities. At one extreme, Genesis, or some other creation story, is taken as literal truth. The biblical story of course, allows but six days for the creation of the Earth and its biota. The rock record is then compressed into the story of Noah's flood. As science, this view is nothing short of preposterous. It flies in the face of both geology and physics. Radiometric dating has provided a coherent chronology for Earth history—a chronology that conforms remarkably well to the sequence of relative ages determined for rocks before physicists discovered radioactivity. During the past few decades, our interpretation of sedimentary rocks has become highly refined. A sequence of beds grading upward from coarse, pebbly, cross-bedded sandstone to layered mudstone can be shown to the characteristic depositional product of a meandering river. From certain peculiarities, thin, flat beds of rippled sandstone can be seen to have formed

FIGURE 8.2
Layers of glacial sediment laid down in quiet water about two billion years ago. A dark layer and a light layer were deposited every year, in an area near Lake Huron, Ontario, Canada.

through the ebbing of a single tide along the border of an ancient sea. Boulders resting in layered mudstone can be shown to have dropped from melting glacial ice into quiet lakes, where one warm-weather layer of sediment and one cold-weather layer were deposited neatly every year. Salt deposits whose thicknesses exceed the heights of our tallest buildings reside within ancient basins, where their change in composition from bottom to top records well-understood changes in water chemistry, as interior lakes or branches of the sea dried up

173

over vast eons of time. The record of sedimentary rocks is a great patchwork of deposits, many of which bear the unmistakable stamp of the environments where they accumulated. Our detailed knowledge of the fabric and mode of origin of these rocks contributes greatly to the discovery of petroleum. When the creationists make the naive claim that all of this is not what thousands of expert geologists have found it to represent, they libel an entire profession.

Creationism also takes less extreme forms. It can be adapted to a geological scale of time by the claim that species are formed periodically by miraculous creation rather than by speciational origins from other species. In its ultimate extension, this idea represents what might be termed the "Will of Allah" point of view: whatever happens is God's choice. Putting it the other way around, God's choice is whatever happens, and this means that a divinity can always be invoked without the possibility of challenge. Science is by no means incompatible with this notion, but neither does science require the notion and, in fact, cannot even address it. The "Will of Allah" viewpoint is untestable, or irrefutable, and therefore unscientific.

Evolutionary science is thus compelled by its very nature to go its merry way without interference from religion, simply assuming that its procedures are not leading it astray. The geneticist Theodosius Dobzhansky spoke against miraculous creation in a way that probably represents the thinking of most evolutionaries who are also theists:

> Those who choose to believe that God created every biological species separately in the state we observe them but made them in a way calculated to lead us to the conclusion that they are the products of an evolutionary development are obviously not open to argument. All that can be said is that their belief is an implicit blasphemy, for it imputes to God appalling deviousness.[1]

The modern creationist's raw materials for schemes of creation are biblical quotations, not evidence gathered from nature. Needless to say, this places creationists at a severe disadvantage: they are attempting to engage in scientific debate while lacking scientific support for precisely what Genesis says. Their recourse represents a kind of scientific nihilism: by attacking evolutionary science, they seek sympathy for their nonscience.

A frequent claim of creationists is that the fossil record contradicts that concept of evolution. One argument here is that there are no transitional forms between distinctive groups of animals or plants. This is not true, and what is most important is that evolutionists do not need dozens of examples to make their case. Even one or two may fly powerfully in the face of Genesis. From

FIGURE 8.3

A series of shells from the Devonian Hunsrück Shale of Germany, representing the derivation of ammonoids from straight-shelled nautiloids. The illustration in the upper left shows two pieces of a long, conical bactritid shell, one of them the bulbous initial portion. The shell evolved a curved and then a coiled shape, perhaps by steps, but the bulbous initial portion remained conspicuous.

anatomical evidence, it was long ago suggested that birds evolved from dinosaurs. In 1861, the discovery of *Archaeopteryx,* a fossil creature that I have already mentioned, provided concrete evidence of connection. Quoting certain scientists who have claimed *Archaeopteryx* to be fully birdlike, creationists have dismissed this interesting form as meaningless, but they are only telling a half-truth. Other scientists have claimed *Archaeopteryx* to be remarkably dinosaurian—to be a dinosaur with a wishbone and feathers! This disagreement is no embarrassment to evolution. Quite the reverse. It underscores the transitional character of the famous fossil.

Archaeopteryx represents a single intermediate form. Elsewhere, despite the punctuational nature of many transitions, we have available series of forms that more fully represent steps in the origins of certain major groups. The ammonoids, extinct octopuslike animals with shells, were long thought to have evolved from straight-shelled nautiloids, ancestors of the modern pearly nautilus. In fact, from the shape of the earliest portion of the primitive ammonoid shell and from the off-center position of a tube running through the shell, it was some time ago suggested that the ammonoids evolved from a particular group of nautiloids known as the bactritids. It was then gratifying that in 1966 H. K. Erben announced his discovery that a group of specimens from the Devonian of Germany could be assembled into a graded series showing several stages in the evolution from bactritid nautiloids to ammonoids. How creationists might explain this empirical substantiation of the evolutionary prediction remains unanswered. Have we been victims of a Divine practical joke?

Creationists commonly cite outdated lamentations of botanists and paleobotanists that the fossil record of plants fails to support evolution. The simple truth is that much of the plant fossil record is terribly incomplete. Also, there has never been more than a handful of people in the world studying fossil leaves, and in the early days serious errors were made in taxonomic assignments. The result was a confused picture of ancient plant life. As I have noted, the biggest problem, the sudden rise of the flowering plants—Darwin's "abominable mystery"—has been resolved by new fossil evidence. We now have fossils documenting a pattern of early adaptive radiation. Here we have a prime example of how the concept of evolution has been strengthened rather than weakened since the time of Darwin.

Unfortunately, modern creationists are well organized, well funded, and (inevitably) zealous. They write books and stage debates; they quote "respectable scientists" out of context and from outdated literature; they cite apparent contradictions within the corpus of evolutionary theory; and they claim that evolution does not represent a true theory (an irrelevant point; we are concerned with credibility, not categorization). Creationists pretend that scientific debate over the details of evolution calls into question the very idea of evolution, but this is an unfair ruse. As the scientific philosopher Carl Hempel has put it, we must "distinguish what might be called the *story* of evolution from the *theory* of the underlying mechanisms."[2]

Unfortunately, formal debate orchestrated by creationists, by its very nature, turns the methodology of science against the popularization of scientific ideas. Because concepts like evolution gain strong support only through testing of their predictions on many fronts and over many years, acceptance of evolution as a near certainty requires considerable knowledge. Meanwhile, the "horse in the Precambrian" possibility remains. Any one of a myriad single observations could refute evolution. This, even a lay person within a creationist's audience understands implicitly. The evolutionist cannot, in a few minutes, present the uninitiated listener with a compelling case for evolution—with a long and varied list of corroborated predictions and an almost infinite number of unrealized possibilities for refutation. The equivalent of a college course would be required. On the other hand, the creationist may announce contrary "facts," any one of which, if valid, would refute evolution. Hearing several of these and being unable to judge their veracity, the audience is often sympathetic and admits its uncertainty as to which view is correct. This is exactly what the creationist wants, because the question he can then ask is whether the audience does not find such debates useful. Should not both sides of the question be widely aired and debated? This apparent exercise of the American town-meeting spirit glosses over the simple fact that a short debate judged by a nonexpert audience is no vehicle for determining whether creationism is science or whether evolution is a well-grounded scientific concept. The judgment must be made within science, and when it is so made, the overwhelming verdict is that evolution should be accepted, and creationism banished from science.

The creationist movement could be ignored were it not for the threat that it poses in American schools. Creationists have, with partial success, mounted a nationwide political campaign to convince states that in the biology courses of public schools, creation should be taught side-by-side, and on an equal footing, with evolution. The problem here relates to freedom *from* religion and freedom *of* religion. As I have attempted to show, science, by its very nature, must be free from religion, and creationism is pure religion. Public schools must also foster freedom of religion. Teaching of fundamentalist Protestant creationism amounts to the establishment by government of a particular religious view, and this is explicitly forbidden by Article I of the Bill of Rights of the United States Constitution.

The world's religions are richly laden with creation stories. The Shinto religion upholds the emperor of Japan as the descendant of the sun-goddess Amaterasu, and the populace of Japan as the issue of lesser deities. The islands of Japan are seen as having emerged from the womb of the primal woman Izanami. Many creation stories are comparably provincial, and some are indigenous to America (Genesis, of course, is imported). The Navaho story, held in as much respect by many Americans as Genesis is by others, recounts the upward passage of humans through four worlds to the present Fifth World. The First World, where life began, was a dark land inhabited by twelve kinds of insect people. The first man and woman came to life in the Fourth World from a perfect ear of white corn and perfect ear of yellow corn, respectively, through the agency of the White Wind blowing from the East and the Yellow Wind blowing from the West. On instruction from the gods, the insect people populating the Fourth World allowed the life-giving winds to blow between sacred buckskins, where the perfect ears of corn had been laid upon eagle feathers. After several years in the Fourth World, the first humans ascended to the present world through a small hole in the sky.

In a nation dedicated to nondenominational government, neither the Navaho creation story nor any other should be officially displaced by Genesis. A fundamentalist Protestant might regard nonbiblical stories as creation myths, but devout members of other sects must regard Genesis in the same way. All stories of creation represent particular religions. None belongs in the secular world of science education.

The interaction between Darwinian evolution and religion began, of course, immediately with the publication of *On the Origin of Species,* but the nineteenth century was remarkable for its thorough expulsion of religion from science. The century began, as we have seen, with religious dogma on the perfection of nature nurturing both a general denial of variability within species and a widespread belief in a *Scala Naturae* and plenitude. Until Darwin applied Malthusian concepts about human populations to animals and plants, it was even widely assumed that, from generation to generation, a species produced only enough individuals to maintain its allotted place in nature. The wastage required for selection was largely unrecognized. Even more remarkable was the effect of the Platonic ideal, adhered to by creationists, which blinded natural science to the existence of variation among the individuals of a species. After suffering this kind of domination, it is remarkable that by the end of the

Continuity or Creation?

century, science had been exalted to a position beyond the reach of religion.

Darwin was not solely responsible for the extrication of science from religious dogma. The Age of Reason, which preceded the nineteenth century, set the intellectual tone for parturition, and long before Darwin, biblical exegeses and studies of Middle Eastern history were casting doubt upon the literal truth of scripture. As early as 1852, the Duke of Argyll, accepting the chancellorship of St. Andrews University, lectured:

> An absolute separation has been declared between science and religion; and the theologian and the philosopher have entered into a sort of tacit agreement that each is to be left and unimpeded by the other within his own walk and province.[3]

Argyll decried the new trend, as did many others in the middle part of the century, including, of course, many of Darwin's opponents.

On the Origin of Species created a watershed, however. Wounded by the new attack on creation, Christianity tended to shrink from any territory claimed by science. In his "Essay of 1844," Darwin complained:

> It is derogatory that the Creator of countless Universes should have made by individual acts of His will the myriads of creeping parasites and worms . . . that a group of animals should have been formed to lay their eggs in the bowels and flesh of other sensitive beings; that some animals should live by and even delight in cruelty. . . .[4]

Christianity responded to the inroads of brazen comments like these from respected scientists with a liberal movement that no longer tied its tenets to the literal truth of the Bible. Christianity became in many ways solicitous, rather than proprietary, in its orientation toward science.

Given the credibility earned by its resiliency under the test of time, evolution will weather the political storms now being fomented by creationists. As genetics and embryology shed more light upon quantum speciation—perhaps in part by the appearance of remarkable new animals and plants in the laboratory—the punctuational model will strengthen, not weaken, evolution in its ascendancy over creationism. The patterns that we recognize in the fossil record are even now becoming increasingly understood through the study of modern life.

Macroevolution and the Direction of Change

SINCE the time of Darwin, humans have often been guilty of taking themselves to represent exactly what evolution should have achieved, to be the inevitable result of a perfecting process of natural selection. This attitude grew out of the older idea of a human-centered universe, and represents the ultimate statement that evolution is progressive. To assume that evolution has necessarily progressed toward a humanlike condition ascribes to nature an evolutionary purpose. Such teleological thinking has not been restricted to scientific amateurs. After the turn of the century, during the period of widespread disillusionment with Darwinism, the concept of orthogenesis gained a modest scientific following. Orthogenesis, an idea founded on a superficial reading of the fossil record, asserts that evolution proceeds rigidly in straight lines. An extreme variation of orthogenesis moves beyond simple description to the antiselectionist notion that some unspecified force actually drives evolutionary lineages on a beeline to extinction.

The gradualistic view of the Modern Synthesis has long since denied evolu-

Macroevolution and the Direction of Change

tionary prescriptions, whether progressive or regressive, and it has debunked even descriptive orthogenesis. Natural selection operates upon variation, and variation arises by accidents of mutation and genetic reshuffling. Who is to say exactly what variations will appear or in what order they will become available for selection? Furthermore, who is to say that the environment will impose selection pressure in one direction for very long?

The punctuational view accentuates the unpredictability of large-scale evolution. Because we know ourselves well, human ancestry again provides a graphic example. I have already noted that there is no simple trend in the evolution of the human skull: a relatively weak-browed australopithecine gave rise to the heavy-browed *Homo erectus,* and by an uncertain pathway from *Homo erectus* came weak-browed modern humans. Our species' high-vaulted forehead—the possible key to our extraordinary brainpower—seems also to have come into being rapidly and unpredictably.

These human examples are not isolated cases within the animal kingdom. In fact, we can generalize. There is much about speciation that is haphazard. When and where will it happen? What population will take part? Which of many possible environmental conditions will obtain? Within the speciating population what kind of variability will erupt for selection to screen? Answers to such questions could never be anticipated by an observer waiting for the next speciation event to issue from some group of species. In other words, a particular speciation event may represent change that is adaptive for a particular place and time—change wrought by natural selection—yet when we stand back and view an entire family or order, we see that at any time, the direction of the next event of speciation will be heavily dependent upon unpredictable historical and genetic accidents.

The gradualistic Modern Synthesis accorded speciation no particularly important role: in the words of Julian Huxley, "a large fraction of it is in a sense accident, a biological luxury, without bearing on the major and continuing trends of the evolutionary process."[1] In the framework of punctuational evolution, things are quite different. The unpredictability of speciation represents an unsettling condition—unsettling because in the punctuational scheme speciation is seen as the focus of evolutionary change. Because of its haphazard quality, speciation represents a kind of experimentation, but experimentation without a plan. It is very much a trial and error process. By speciation, orders and families of animals and plants continually, but unknowingly, probe the environment.

What happens when we view evolution within a family or order for a long period, as we do after the fact by studying the fossil record? We then see that major changes have occurred within large groups of animals and plants, and we find that these changes have often been sustained and directional. They are what we call evolutionary trends. Modern gradualists have interpreted large-scale trends in the way that Darwin did—as changes representing the gradual remodeling of species over vast stretches of time. If this does not happen—if species are remodeled rather little before dying out, and most change is concentrated in speciation events of unpredictable character—what accounts for large-scale trends? What brings order out of the chaos of haphazard speciation? How, for example, did four-toed, leaf-browsing horses give way to the prevalence of single-toed grass-grazers that we see in Ice Age deposits and in the modern world? A major part of the answer is that some speciation events have been more successful than others. They have produced certain species that have happened to fare well. What does it mean for a species to fare well?

The origin of a particular species in a particular place represents immediate success, but this is no guarantee of good fortune in a broader environmental context. What I mean by success of a species is the displaying of one or both of two traits: first, survival for a long time, and, second, rapid production of descendant species that share the species' basic attributes. These two traits form my definition of success for a particular reason, which can be understood with reference to figure 9.1. This is a diagram of a segment of the tree of life—an order, or perhaps a family, of animals or plants. The diagram depicts a net trend, in that the average shape of species within the group changes through time, and it changes rather persistently during the interval considered. By what mechanism has the trend been wrought?

First, let me say what has not contributed to the figured trend. The trend has not developed by evolution within established species, because the diagram exaggerates what happens in the real world by portraying no evolution whatever within established species. Nor has the trend developed by a tendency for most speciation events to move toward the right or to take larger steps toward the right—the direction taken by the trend. In fact, the diagram has been carefully arranged so that the same number of speciation events move to the left as to the right, and the total amount of change contributed by all speciation events moving to the left equals the total amount contributed by all events moving to the right. What then produces the trend? At this point, the question may seem like a riddle, but I have already supplied a strong hint.

The trend results from the simple fact that species positioned toward the right side of the diagram tend to produce more descendant species than those positioned toward the left side. This results in a shift in the average shape of

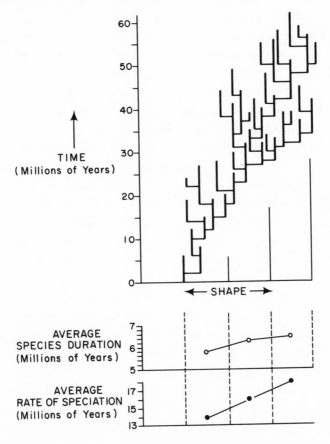

FIGURE 9.1

A hypothetical evolutionary trend (change in shape) produced by species selection within a branch of the tree of life. Here a single unspecified shape variable is plotted on the horizontal axis.

species toward the condition represented on the right side of the diagram. There are actually two conditions that increase the relative numbers of species positioned toward the right. One is that species positioned toward the right survive longer than those positioned toward the left. Just as long-lived individuals tend to produce more offspring than short-lived individuals, species of great geological duration tend to produce more descendant species than species whose stay on earth is brief.

The second condition contributing to the trend is equivalent to differential fecundity among individuals within a population. This condition is a differential rate of production of new species; quite apart from their longevity, species positioned toward the right side of the diagram produce more descendant species per unit time (per million years, for example) than species positioned toward the left.

The analogy between successful individuals and successful species is of great importance. Individuals that produce a disproportionate fraction of offspring are the kind favored in natural selection within a species. What is depicted in the diagram showing the large-scale evolutionary trend is an analogous form of selection that operates at a higher level—selection among species, or what I simply call species selection.

The concepts I have outlined are sufficiently complicated for a summary to be in order. The point of departure is the fact that while well-established species do evolve by natural selection, their evolution is too sluggish to produce most major evolutionary trends (net changes in the properties of the tree of life over long stretches of geological time). Natural selection may be responsible for most of the change that occurs during speciation as well, but its direction is determined in each speciation event by the kinds of chance factors identified earlier. These chance factors influence the type of environment where speciation happens to occur (a matter of geographical and geological accident), the initial character of the speciating population, and the kind of variability that happens to arise within the population. All of this means that we cannot predict with any degree of certainty the direction of the next speciation event within any segment of the tree of life. In the somewhat exaggerated case depicted in the diagram, equal numbers of events move in each direction along the particular axis that has been chosen. In this hypothetical case, we have no predictability whatever. Each event is like the flipping of a coin. In nature, there is more bias in the direction taken by speciation, but the odds against predicting the direction of the next speciation event are nonetheless very high, particularly when all possible directions of change are considered. The smart money should find another game.

Because the direction of speciation is highly unpredictable (or, we might say, highly random), macroevolution (change in the average biological condition of a family or other large group) is decoupled, or disengaged, from microevolution, or change within a single population or species. Not only is unpredictability in the direction of speciation a factor in my distinguishing between macro-evolution and microevolution, but the role of extinction is of special importance. Here a concept that has been labeled "emergence" enters in.

Emergence pertains to the study of natural systems at more than one level.

Macroevolution and the Direction of Change

Emergent qualities are those imputed to appear uniquely at a level above the lowest and to require investigation at that higher level if certain phenomena are to be understood. In its most extreme form, the reductionist view of science denies altogether the presence of emergent qualities, asserting that study of a system at the most elemental level will permit us to learn all that we have the ability to learn.

With regard to the life sciences, we live in the era of molecular biology. In this field, the popular reductionist view holds that living systems can be reduced to chemistry, or, ultimately, to particle physics. According to this idea, even the study of natural selection among the individuals of a population is a superficial business, ultimately to be replaced by investigations of molecules, atoms, or subatomic particles.

My special concern has been with analysis at a still higher level than selection within a population or species—at the level where species, rather than individuals, are the units that multiply and die out at varying rates. Our question, then, is whether emergent qualities exist at the level of the species. The died-in-the-wool reductionist would dismiss this issue, claiming that both levels will eventually be replaced by inquiry at more elemental levels. I intend to skirt this thorny ultimate question, but cannot resist pointing out that the reductionist's mission is complicated enormously by the necessity that not only organisms themselves be understood, but also environmental change. Environmental change entails vagaries of geological and even astronomical events whose origins are complex, obscure, and, as earthquake watchers know, currently unpredictable.

Certainly for now, it is fruitful to evaluate change within a population by employing the individual as a unit that engages in reproduction. It would be an idealistic and futile exercise to abandon study at this level, with all that it has to offer, simply to satisfy a purist opinion that such analysis will eventually be supplanted by more elegant and comprehensive studies at a lower level. Even the philosophical reductionist must recognize that science peels away layer after layer in approaching a core of understanding. Practicing scientists must operate at those levels that, during their particular lifetimes, bear fruit.

Let me, then, leave to philosophers the question of ultimate reductionism for biological systems and focus instead on analysis at two higher, organismal levels: the level at which selection occurs among individuals (microevolution) and the level at which selection occurs among species (macroevolution). Are there emergent properties that justify a separate but equal status for macroevolution?

One of the two components of species selection is the operation of differential rates of extinction among species. Does extinction fall within the domain of natural selection? The answer is "no." Extinction is simply the death of all individuals within a species. It is the particular case in which all kinds of individuals fail to survive and reproduce. Natural selection, on the other hand, is the process whereby some individuals survive and reproduce more successfully than others. For selection to be measured, some individuals must survive! Extinction is an aspect of demography, or of population dynamics analyzed in simple numerical terms, not an aspect of evolution within a population. (In fact, we can view extinction as the numerical result of a failure of evolution to provide the adaptive change adequate for survival.) Here, then, is a way in which the study of natural selection fails to confront macroevolutionary change—change in the biological character of a related group of species, change that is partly wrought by differential extinction. With respect to the explanatory power of microevolutionary processes, the presence of extinction represents an emergent property of macroevolution relative to microevolution. This property, if nothing else, validates the concept of macroevolution.

Another argument relates not to emergence, but to practicability. It resembles, in principle, the idea that even biologists of a reductionist bent who wish to study evolution within populations are currently unable to answer evolutionary questions with only chemistry and physics in their tool kit. Whether the reductionist philosophy is valid or not, our technological prowess and scientific sophistication at present are simply inadequate to the reductionist's task. If we are to study macroevolution, including the ancestry of our own species, we must employ information from the fossil record. Unfortunately, this limits us to fragmentary evidence. The fossil record is far too incomplete to allow us to come to an understanding of most details of selection within ancient populations. In general, we do not even possess a good enough record of ancient species to reconstruct complete patterns of species selection, but at this level we can at least draw some inferences. We can rough out patterns of diversification and decline, and sometimes, as I will describe shortly, we can identify agents of change.

The concept of species selection has many important consequences. At the very beginning of this book, I observed that there has been a traditional idea that major evolutionary transformations are brought about by slow, persistent changes in the environment. "What drove human ancestors down from the trees?" has been the question. "What forced horses to change slowly from a habit of browsing on leaves to one of grazing on grass?" In the punctuational view, such queries lose much of their significance. Major changes can take place

Macroevolution and the Direction of Change

even without environmental trends, simply because new species are always forming and some of them inevitably happen to end up being highly successful. Speciation provides for rapid evolutionary forays into uninhabited ecospace. In this sense, evolution becomes highly opportunistic. Having attributes that serve them well, some new species tend to survive unusually long or give birth to descendant species at a high rate. A macroevolutionary trend may result—a trend toward greater numbers of such species and often, in time, toward other species that have even more pronounced features of the same type. Certainly, species selection is sometimes guided by long-term environmental change, but the point is that the species itself does not respond very effectively by transformation. It survives with modest change or dies out. Rather, it is the *group* of species that responds most effectively—the genus, family, or order—by changing its composition through species selection. Sometimes, of course, the kinds of species appropriate for new conditions fail to appear within the group. Then species selection fails and the group becomes extinct.

What brings about selection among species? Obviously many agents of species selection are environmental in nature. These agents must be those aspects of the environment that determine how long various kinds of species survive and how successful their populations are at blossoming into similar new species. The environmental conditions must, then, be the very ones that control the distribution and abundance of populations and species: they are what ecologists call "limiting factors" of the environment. These are: predation inflicted by other species; competition from other species for such things as food and space; the impact of environmental controls such as climate and food supply; and (for small populations) chance factors such as accidents of reproductive success.

While environmental agents seldom drive evolution very far within large, well-established species, they must often dictate the extinction of species (the ultimate limitation of distribution and abundance!). They must also frequently determine whether or not small populations blossom into new species and where and when speciation occurs. Thus, limiting factors give populations and species life, and they take it away. This is how they guide species selection.

Let me offer one example, in which competition seems to be the dominant limiting factor. Perhaps being biased by his own interest in barnacles, but with some justification, Darwin branded the modern segment of geological time the "age of barnacles." It is true that barnacles are found in many places, including most rocky shores and the hides of many whales, but Darwin's edict nonetheless needs amending. It is not the age of *all* barnacles. On rocky shores around the world, one group, the balanoid acorn barnacles, are coming into ascendancy, while another group, the chthamaloid acorn barnacles, are on the decline. The balanoids, which are the younger of the two groups, seem to have squeezed most intertidal populations of chthalamoids into the uppermost reaches of the intertidal zones of rocky shores, where balanoids are unable to live. Competition for space is often keen among the many forms of marine life that attach to rocky intertidal surfaces. The chthamaloids were apparently doing quite well until the balanoids evolved about forty million years ago (not far back in geological time), but when balanoid evolution got underway, it was with a special advantage. This was that the balanoid external skeleton was full of hollow places. As with human bones and concrete blocks, little strength is lost with this kind of construction, but, being built with little material, the porous skeleton of balanoids offers the advantage of growing rapidly while remaining firmly attached. As a result, balanoid barnacles now tend to crowd out chthamaloids on rocky surfaces, overgrowing them, crushing them, or breaking them from their moorrings. Whereas most solid-walled groups are on the decline, apparently having a high rate of extinction and a low rate of speciation, balanoids and members of another hollow-walled group are speciating rampantly. In other words, species selection is favoring barnacles with rapidly growing, hollow skeletons. This kind of skeleton has apparently evolved more than once during the recent history of the acorn barnacles, and it is an example of an evolutionary experiment that has happened to be of great adaptive value.

Not all groups that win in species selection win because of adaptive superiority, and not all are blessed with both a high rate of speciation and a low rate of extinction. Some triumph simply by virtue of an inherent tendency to speciate at a high rate. The cichlid fishes, for example, appear to undergo quantum speciation more readily than do other groups of fishes. In chapter 6, I noted that this seems to relate to the ease with which their jaws and teeth are extensively modified by simple genetic changes. The flowering plants, discussed in chapter 5, may have overwhelmed preexisting plant groups in large part simply because their pollination by insects and birds offers a special mechanism for frequent speciation: reproductive isolation can be effected easily by a change of animal pollinators. Obviously, the diversification of flower structures and the diversification of pollinators have gone hand in hand. On a smaller scale, we can note that flowering plants vary in their modes of association with pollinators. Those plants whose pollination systems have offered the greatest likelihood for the reproductive isolation of small populations by association with new pollinators—for speciation—may often have become the richest in number

Macroevolution and the Direction of Change

of species: they will have been favored in species selection. The pollinators associated with these groups will have benefited in reciprocal fashion.

I could offer other apparent examples of species selection, but this seems unnecessary. Let us, instead, consider several important implications of the process and, more generally, of the opportunistic nature of punctuational evolution. One of these implications relates to the idea of optimization. The gradualistic view has seen evolution as a rather efficient perfecting mechanism. In the era of the Modern Synthesis, it has been widely assumed that the species remains finely tuned to its environment, meeting external change with evolutionary adjustments and thus tending to optimize its adaptive relationship to its milieu. The assumption has been that when the environment dictates a change, appropriate variability is present within a species for effective response. The punctuational view sees things otherwise. It allots the species only restricted flexibility. The average condition of a family, order, or other large group may shift by species selection, but the evolution of an individual does not track major environmental changes effectively. What better indication of this inflexibility could be found than the evidence embodied in living fossils, the stable forms that, I have claimed, illustrate that speciation is required for substantial evolution to take place. This is a very different notion than Darwin held. Optimization was at the heart of his world view.

Through such historical considerations, we confront again another idea of Darwin and his contemporaries—the idea of plenitude, or the fullness of nature. As we saw in chapter 3, Darwin retained from the *Scala Naturae,* which his evolutionary theory eclipsed, the idea that the world is full of species which, in an ecological sense, stand shoulder to shoulder, niche against niche. I have shown that this notion was one of the things that prevented Darwin from envisioning punctuational evolution. In his system, a species could only be transformed if its neighboring species made room. The nature of optimization for a species was determined largely by its competitive boundaries. Little variation was to be expected with a species—and only slow evolutionary change, which was seen as necessarily moving in close concert with changes in other species. To Darwin and to most framers of the Modern Synthesis, the structures of established species formed a highly dynamic, interactive system. To the punctuationalist, the system as a whole is dynamic, but its components—species, the units of large-scale change—are less plastic. Changes come mainly by the addition and subtraction of these units, not by their remodeling.

Extinction leaves voids in nature, possibilities for life where none exists. Given these voids, the punctuational nature of evolution further militates

FIGURE 9.2

Reconstruction of *Indrichotherium*, perhaps the largest mammal of all time. This herbivorous giant lived in mid-Cenozoic times.

against perfect plenitude. Neither the sluggish modification of existing species envisioned in the punctuational model nor the haphazard production of new species provides an efficient mechanism for maintenance of plenitude. (Of course, as we have seen, speciation is often accelerated after major extinctions, but this is the extreme case and by no means guarantees that complete saturation will be sustained.)

We can document the absence of plenitude in many regions of the modern world. Modern continents are populated by an abnormal fauna of quadrupeds. The elephant, rhino, and hippo seem gigantic to us today, but many kinds of extinct mammals have equaled or exceeded them in height and weight. Once the imperial mammoth, a larger beast than its modern elephant cousins, roamed the New World. There has also been a North American bison so large that its horns spread to more than six feet. Among the most conspicuous South American behemoths have been the twenty-foot species of ground sloth and the nine-foot species of armadillo excavated by Darwin. The world has seen rodents the size of hippos; flightless, predatory birds nearly as tall as a man; and camels, deer, wolves, and lions much larger than any alive today. The some-

Macroevolution and the Direction of Change

what impoverished nature of modern faunas of land animals tells us that terrestrial habitats today are not brimfull with life.

Today, in some regions, the marine realm is also biologically depauperate compared to its past condition. Germaine L. Warmke and R. Tucker Abbott wrote a book, *Caribbean Seashells*,[2] in which they described and illustrated about a thousand living species—most of the shelled mollusks living in the Caribbean. We think of Caribbean seas as teeming with beautiful shelled mol-

FIGURE 9.3
Skeleton of *Mesembriornis*, one of the giant, flightless birds that once inhabited South America. The golden eagle shown beneath its head serves as a scale. *Mesembriornis* lived about five million years ago.

FIGURE 9.4

Megaloceros, the Irish "elk" of the Ice Age. It was really a giant deer, and its antlers spread up to thirteen feet across.

lusks, yet our perspective is again biased. The fossil record of the Caribbean shows that before the Ice Age this region was populated by many more species than exist there today—perhaps two or three times more! During the Ice Age, most of these species became extinct, apparently by reduction of temperatures when at high latitudes great volumes of water were locked frozen upon the land. The existing molluscan fauna by no means saturates Caribbean environments: there is no question here of plenitude.

Part of the reason that plenitude does not actually obtain in nature is that species do not evolve effectively, but there is another reason as well. This has to do with a very important limiting factor: predation, or the feeding of some species upon others. A predator whose populations are not themselves held in check can decimate populations of its prey. From this simple fact comes a primary principle of ecology. To appreciate this principle, let us return to the rocky shore, the scene of internecine war among modern acorn barnacles. Some

Macroevolution and the Direction of Change

stretches of intertidal rocky shore are not, in fact, the sites of severe competition. These are often stretches characterized by intense predation upon the several species that attach to the rocks. When predators (starfishes and carnivorous snails, for example) are artificially removed from a section of shore, the community of organisms changes. What happens is that one species of shore animal, sometimes an attached mussel, overwhelms all others, multiplying greatly and monopolizing the rocky surface. This kind of simple predator-removal experiment shows that when the potential victors in competition for space nor-

FIGURE 9.5

Mollusks that flourished in South Florida until the Ice Age. A: The cone shell *Conus* (*Leptoconus*) *waccamawensis*. B: *Trigonostoma sericea*. C: *Vasum horridum*. D: *Trachycardium emmonsi*. E: *Cancellaria amoena*.

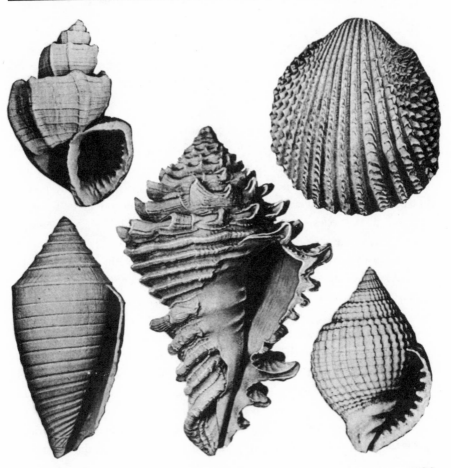

mally have their numbers held in check by predation, many species can coexist; there is plenty of space for everyone, and, at almost any time, space is left over for new colonization. In many other ecosystems this is also the normal state of things; voracious predators are often present, suppressing populations of their prey to leave an excess of space and food for other species.

Strangely, the notion that predation can affect an increase in the diversity of species able to live together came to the fore in ecology only during the 1960s, as a result of the experimental observations of Robert T. Paine of the University of Washington. Even stranger, perhaps, is the fact that Darwin failed to incorporate this idea into his evolutionary theory, clinging instead to the idea that species are tightly interlocked in competitive interactions under conditions of plenitude. I label this as strange because, little known to many modern ecologists, Darwin himself enunciated the predation principle in *On the Origin of Species*! Here he reported on his personal experiments with plants (p. 67—68):

> If turf which has long been mown, and the case would be the same with turf closely browsed by quadrupeds, be let to grow, the more vigorous plants gradually kill the less vigorous, though fully grown, plants: thus out of twenty species growing on a little plot of turf (three feet by four) nine species perished from the other species being allowed to grow up freely.

Perhaps because of the firm entrenchment of the concept of plenitude, Darwin failed to grasp the broad implications of his principle. The *Origin* is imbued with references to fierce competition among species in nature, with little recognition that such competition might be alleviated among heavily preyed-upon species, and that at any time there must be room on Earth for many more of these species than actually exist. (Recall Darwin's analogy between species and wedges.)

I can easily envision only one historical possibility for the reality of global plenitude. The interval of time I have in mind is the long stretch of the Precambrian that preceded the appearance of the first cell-eating creatures. During this interval, ecosystems were very unusual by modern standards. There were no multilayered food webs: there were no multicellular creatures, only aquatic algae, and these were, therefore, not eaten by animals (or by single-celled animallike creatures). Instead, in this world of primordial slime, algae and bacteria had aquatic habitats essentially to themselves, and the land was essentially barren.

When we reflect upon the simple, primitive ecosystem I have just described, our attention must be drawn to the predation principle that I outlined earlier—

Macroevolution and the Direction of Change

FIGURE 9.6

A two-billion-year-old sea floor covered by stromatolites long before destructive animals had come into existence. A geologist's hammer shows the scale. Stromatolites are layered, mound-like structures built by algae that trap and bind sand and mud. This particular stromatolitic surface was exposed in northern Canada by the scouring activities of Ice Age glaciers.

the principle that predation makes room in the world today for many species by holding in check the potential winners in competitive interactions. The early Precambrian world was one in which there were no predators performing such a role. We must imagine that algae floating in oceans and lakes and growing on sea floors and lake bottoms multiplied to the limits determined by the supply of environmental resources—the supply of sunlight, for example, and of compounds providing essential nitrogen and phosophorus. We must imagine that aquatic habitats were saturated with algae—that plenitude reigned.

For the sea floor, the rock record provides direct evidence of plenitude, or a condition that approached it. The Precambrian record reveals that vast stretches of the ocean bottom were blanketed with what are called stromatolites—mound-shaped structures formed by mats of single-celled blue-green algae that alternate with layers of sand and mud. Blue-green algae are more like bacteria than like higher algae. Their cells share with bacterial cells such primitive features as the absence of a nucleus with discrete chromosomes. From a time well back in the Precambrian until early in the Phanerozoic, when marine animals became abundant, stromatolites formed by blue-green algae were extremely abundant and widespread. They declined only with the initial adaptive radiation of grazing and burrowing animals—animals that destroy algal mats and prevent them from flourishing sufficiently well to form stromatolites. Interestingly, the kinds of blue-green algae that can form stromatolites exist throughout the world today, but they are able to produce stromatolites only in hostile environments, such as the exceptionally salty Shark Bay in Australia, from which potentially destructive animals are excluded.

It is a remarkable fact that the kind of algal ecosystem in which plenitude may have obtained existed for most of the Earth's history, in fact, for the better part of three billion years. The great adaptive radiation of marine animals and of those animallike single-celled creatures that have left fossil remains was not underway until less than a billion years ago.

It seems to me that the ecological impact of predation may provide an explanation, or a partial one, for the long delay in the origin of animals and animallike creatures. I believe that before there were effective predators, algal systems were more-or-less saturated. This means that whatever species existed could not easily have given rise to new species. Had they existed, consumers (single-cell predators and multicellular animals) could have relieved resource limitation to accelerate speciation and, hence, evolution in general. The problem was that speciation was necessary for the very origin of consumers! This amounts to a "Catch-22." We can envision a saturated world in which only occasional speciation events were tolerated. Thus, for nearly three billion years, evolution was very sluggish. Finally, through the agency of infrequent speciation events, the first single-celled creatures capable of eating other cells evolved. This breakthrough initiated the relief of resource limitation and perhaps allowed for new kinds of plant evolution. Most importantly, it inaugurated a new kind of ecosystem—the kind that we know today, the kind that includes not only plants but animals. As we have seen, according to the punctuational model it was inevitable that the appearance of the very first animallike creatures would be followed by an adaptive radiation, such as the sort we know did indeed usher in the Phanerozoic interval of Earth history.

There is one kind of modern ecosystem in which we may find an analogy with Precambrian plenitude. This is the so-called eutrophic, or "overfed," algal system of lakes that are polluted by nutrients supplied at an unnaturally high rate by human culture. Here, indeed, we often find one or a few varieties of blue-green algae monopolizing the available resources and thriving to the exclusion of other algae or higher plants. Often animals do not relieve the saturated condition of eutrophic systems because the decay of dead algae depletes the water of oxygen. Here, in the absence of predation, a kind of artificial plenitude reigns.

Reference is frequently made to the "balance of nature." It is true that the presence of a good many of the species of the modern world is linked to the presence of others. To consider an obvious example, parasites are utterly dependent upon their hosts. Still, to suggest that all species in any habitat are interdependent to such a degree that we must fear the domino effect—the loss of species after species when one is removed by extinction—is to exaggerate grossly. I have noted that Darwin clung to gradualism in part because he believed species to be so tightly wedged together that the environment could normally impose selective change only by the passage of gentle waves of disturbance from species to species. My point that the extinction of a single species does not always have a domino effect represents a complementary idea to my earlier point that at any time there is much room in the world for quantum speciation.

The idea of widespread plenitude was perpetuated well into the present century by ecological experimentation and theoretical analyses focusing on competition between species, as if this prevailed everywhere. Species was pitted against species in the laboratory, and mathematical equations were concocted to assess the likely outcomes of species interactions. While we now know that competition is often important, there are many ecosystems to which species can be added and from which they can be removed with little impact upon others. These are usually systems in which predation, disturbance of the environment, or other sources of mortality hold populations at sparse levels. This point we have generally recognized only since the 1960s.

I must grudgingly confess to harboring disagreement when environmentalists claim that every species, however inconspicuous, plays a crucial role in the balance of nature. We simply cannot uphold some species as essential cogs in the machinery of their ecosystems. We can nonetheless value all species through our aesthetic sensibilities, from a belief in the sanctity of life, and for the potential contribution of each species to our understanding of nature. The lessons that endangered species of Death Valley pupfishes may teach about speciation form but one example (chapter 6).

Just as fundamental as what the punctuational scheme implies about evolutionary trends, optimization, and plenitude is what it suggests about sexual reproduction. It perhaps seems improbable that such a connection would exist. In fact, the connection is quite strong and not at all unwelcome, because during the reign of gradualism, sexual reproduction has been enormously problematical. The question has been, why do sexual modes of reproduction prevail in the animal and plant kingdoms?

The traditional, gradualistic idea has been that sex functions to accelerate evolution within species: it benefits individuals, or species, or both, by continually shuffling genetic material so that new genetic combinations are available to meet environmental change. Here again we encounter the gradualistic notion that, in order to survive, species must evolve effectively in concert with what goes on around them. With regard to sex, the idea, then, is that asexual creatures—those that produce only clones—cannot evolve effectively enough to survive for long. They appear from time to time but are ephemeral and few are to be found in the world at any time. There are many possible variations among theoretical geneticists as to whether any or all may be valid. It is fair to say that gradualism has been quite uncomfortable with sex.

The punctuational theory offers quite a different perspective. Its tenet that established species do not actually evolve effectively casts doubt upon the idea that species must possess sexual reproduction to do just that. Rather, speciation is where the action is, and this is where we must look for the central role of sex. According to nearly all views, genetic recombination in some form is crucial to the process of quantum speciation: new body plans and behavioral patterns must develop rapidly, with genetic rearrangements and re-pairings playing an important role. It seems safe to say that divergent speciation would be difficult without genetic recombination. This means that divergent speciation is difficult for sexless creatures. Here we arrive at a punctuational argument for the prevalence of sex. Without sex there is no real speciation, only the very gradual divergence of clonal lines of descent, as these accumulate occasional mutations. This is a perilous situation. In the perspective of geological time, we must view all species as ephemeral entities. If a distinctive new family of animals or plants exists early on as a single species, that family will not survive for long unless that solitary species gives rise to others. The family must diversify to survive, and an asexual clone, having little or no ability to undergo rapidly divergent speciation, is not likely to last long.

Asexual forms crop up occasionally in the course of evolution, as a part of the variety of life engendered by speciation. (They reproduce in various ways, but in general can be viewed as females that do not require fertilization of their eggs.) The point is that these asexual "experiments" do not in general fare well in the great game of species selection. In fact, one could say that they do not fully participate because they do not actually speciate. Unfortunately for them, they do, nonetheless, participate in the negative aspect of species selection: like sexual species, they are subject to extinction.

Macroevolution and the Direction of Change

In the world today, there are very few large groups of asexual animals or plants. The bdelloid rotifers constitute the largest. These are simple, tiny "wheel animals" that live in moist places, for example, among the fronds of mosses or in tiny pools of water at the bases of the leaves of flowering plants. What is the bdelloids' secret? It is almost beyond belief. If they dry up, the recipe for bringing them back to life is simple: add water. If frozen, they recover fully when thawed. It is difficult to imagine what might cause extinction within such a hardy group, and this is the key point. Their rate of extinction is apparently so low that there is no need for it to be offset by divergent speciation. For this remarkably resilient group, slow clonal rates of diversification seem to suffice.

Ideas of selection among species have a long history. Struggle among species was firmly in the mind of the geologist Lyell, whose *Principles of Geology* profoundly influenced Darwin in the 1830s. In that great work, Lyell wrote:

> In the universal struggle for existence, the right of the strongest eventually prevails; and the strength and durability of a race depends mainly on its prolificness. . . .[3]

By "race," Lyell meant "species." He was addressing "prolificness" within species, but his ultimate concern was with the failure of species that were engaged in the struggle. Lyell was not an evolutionist at the time, and he avoided the question of species' origins, so that he was not fully entertaining the idea of species selection, but he was considering one of its facets: differential success against extinction.

Sandra Herbert[4] has pointed out that Darwin's conception of natural selection came with his transfer of the struggle between species, as envisioned by Lyell, to a struggle between individuals within species. This was triggered by Darwin's reading of Malthus, which convinced him that many more individuals were born within a species than could possibly survive. (Previously he, like Lyell, had assumed that a species produced only enough offspring to sustain its established position in nature.)

The evidence that Darwin conceived of evolution by focusing the struggle for survival and replication at the level of the individual brings a certain irony to the punctuational claim that large-scale change requires analysis of selection at the level of the species. In fact, Darwin himself continued to recognize selection among species. This we can see, for example, in a passage in the *Origin* (p. 130) from which I quoted earlier where Darwin speaks of selection within the tree of life operating "as [within a real tree] buds give rise by growth to fresh buds, and these, if vigorous, branch out and overtop on all sides many a feebler branch. . . . '' The branches are evolving species. In fact, the subtitle of the *Origin, On the Preservation of Favoured Races in the Struggle for Life,* speaks of differential extinction of species (the entities Darwin meant by "races"). It is nonetheless evident that, while Darwin did not restrict selection to the level of the individual, this is where he saw it as being focused. In the *Origin* (p. 75) he wrote, "But the struggle almost invariably will be most severe between the individuals of the same species. . . . ''

While Darwin's emphasis was upon selection within species, for a number of reasons he actually blurred the distinction between this level of selection and selection between species. One problem was that Darwin had an uncertain notion of the species as an entity. A profound effect of his evolutionary theory was that it connected together what had been seen as fully discrete species within the *Scala Naturae*. He then reasoned that if species arise gradually from varieties of other species, they must exist today in varying stages of development. To Darwin, the species, then, lost its identity as a natural biological unit (*Origin*, p. 52):

> From these remarks it will be seen that I look at the term species, as one arbitrarily given for the sake of convenience to a set of individuals closely resembling each other, and that it does not essentially differ from the term variety, which is given to less distinct and more fluctuating forms. The term variety, again, in comparison with mere individual differences, is also applied arbitrarily, and for mere convenience sake.

To what extent Darwin, when composing the *Origin*, believed that species existing at one time are reproductively isolated entities has been a matter of debate. The biological historian Garland Allen has written: "Darwin's nominalist position was essentially mechanistic—that species do not exist, what existed were only individuals interacting with one another."[5]

Darwin's belief in a continuum between the individual and the species extended to still higher categories. It was his gradualistic view that natural selection slowly molded an interbreeding population into a variety and eventually, barring extinction, into a genus or family. Belief in this gradual progression prevented Darwin from isolating species selection as a totally separate process. In all these matters it appears that Darwin's replacement of the compartmentalized *Scala Naturae* with gradualistic evolution imparted a momentum that moved him past the notion that species are entities—naturally interbreeding groups—that could themselves be the units of a discrete selection process.

Darwin's emphasis upon selection within species was underscored in the gradualism of the Modern Synthesis, which for several decades suppressed the

FIGURE 9.7

Wolflike marsupial mammals. *Prothylacyrus* (A), from the Miocene of Argentina, though smaller, closely resembles *Thylacinus*. (B), a genus that evolved quite independently in Australia. *Thylacinus* became extinct only after humans colonized that continent.

notion of selection among species. It has only been with the recognition that many of the discontinuities that exist between species have developed rapidly (albeit usually by selection among individuals) that the complementary process of species selection has reared its head to take on new meaning as a discrete process. Macroevolution is a valid subdivision of evolution in general, and the vast scope of its temporal dimensions places it squarely in the domain of pale-ontology, where species can be seen to have multiplied and died out at differ-ing rates.

While I have laid great importance to species selection as a process that brings about major evolutionary trends, it is not the only process at work. Speciation is not totally random. Some trends have resulted from a tendency for speciation to move in a particular direction. Human evolution perhaps offers an example. While the human brow has evolved from weak to robust to weak again, hominids seem, step by step, to have evolved larger bodies and brains. Possibly an increase in size of body and brain has been a near necessity for success of new species within the human family. If so, certain important general trends in human evolution may have resulted primarily from a bias in the direction of speciation—from what I have called "directed speciation"—rather than from species selection.

Thus, a stepwise series of speciation events provides one source of evidence for directed speciation. Another line of evidence comes from parallel evolution. Parallel evolution is the evolution of a distinctive kind of animal or plant from another more than one time. The marsupial mammals offer a good example. In both South America and Australia, the Cenozoic adaptive radiation of marsu-pials produced a wolflike predator. The two predators were remarkably similar in form (they are now extinct), and they resembled each other more than they resembled true wolves, which are unrelated placental mammals. The parallel development of the two wolflike marsupial forms on separate continents sug-gests that, given the ecological opportunity, primitive marsupials had a tenden-cy to give rise to predatory species of this particular type. The trends that led to the marsupial "wolves" probably followed similar stepwise sequences of evolu-tion from a generalized ancestor, although in neither case has the trend yet been reconstructed from fossil data.

Serving as a reminder that we must not overplay the role of parallelism and directed speciation, however, is the presence of kangaroos in Australia. The origin of a large kangaroolike mammal must be regarded as improbable and unpredictable in both time and location. As far as we know, no large animal comparable in form and bounding locomotory system has evolved elsewhere in the world. This, and other less dramatic examples of the "experimental" qual-ity of speciation, shows how that differential success among species—species selection—is of great importance in establishing the overall course of evolution.

10

Punctuationalism and Society

I F EVOLUTION in general bears on our image of ourselves in the universe, then the particulars of the punctuational model must mold this image in certain ways not appreciated during the reign of gradualism. Among the potential lessons of punctuational evolution are ones relating to the structure of cultural change and the future evolution of humanity. It seems wise to tread lightly in these areas, and my comments will be brief, though perhaps provocative.

My most fundamental point has been that the evolution of well-established species tends to be very sluggish—too sluggish to account for most macroevolutionary change. This we can read from the fossil record, where we find that the entities recognized as species have spanned long intervals of geological time—much longer intervals than biologists have assumed. Stability of species recalls stability of institutions, and this leads to the question of cultural analogs to evolution. Such analogs usually appear under the rubric "Social Darwinism," but their most eminent investigator, Herbert Spencer, was an evolutionary predecessor, rather than follower, of Darwin. Several years before Darwin went to press with the idea of natural selection, Spencer published selectionist ideas on social evolution. Spencer, however, was also a Lamarckian.

Before Darwin, the Frenchman, Jean Baptiste Lamarck, believed that the use and disuse of organs, through an animal's conscious striving, led to heritable changes in shape. Culture or behavior that develops anew during a person's lifetime can, of course, be passed on to members of subsequent generations. In this kind of inheritance of acquired characteristics and in people's willful adoption of new traits, behavioral evolution does indeed resemble Lamarck's fallacious view of organic evolution.

Despite Spencer's efforts, comparisons between cultural evolution and evolution in nature came into prominence only after Darwin's publication brought widespread attention to biological evolution. A great misconception was then widely promulgated. This was that cultural evolution is a natural extension of biological evolution and should be guided by exactly the same principles. Spencer's initial fixation upon this "natural" ethic related to his concern that traditional theological justifications for ethical absolutes had been undermined by science. To him, it seemed incumbent upon the conquering discipline to provide for replacement: physical and biological laws of nature should supply society with ethical and behavioral guidelines.

It is ironic that the evolutionary analogy for culture rose to great popularity only when natural selection came to be widely accepted. In truth, cultural evolution, being characterized by the inheritance of acquired characters and by a role for conscious choice, draws equally upon Lamarckism and Darwinism. The fact that the central thesis of Lamarckism is invalid means that cultural evolution has no close natural analog. Consequently, it does not demand rigid guidance by a natural Darwinian ethic! Social Darwinism might equally well have been labeled "Social Lamarckism." Its lack of legitimacy on this count was exposed late in the nineteenth century by such men as Lester Ward, an intellectual of broad achievement who for a time served as chief paleontologist of the United States Geological Survey.

There is then a fundamental flaw of logic in the idea that cultural evolution should necessarily be dictated by laws of nature. Culture, by definition, includes behavior and custom that is not coded genetically. This automatically separates it from biological evolution. What the concepts of biological evolution can contribute is instructive analogy that, in our conscious governance of cultural evolution, we can apply to our decision making. Not only do we mold our culture, but we mold the environment in which our culture evolves. This was not recognized by Spencer, who saw the environment as constant, but was emphasized by the philosophical pragmatists, who came into prominence later, after the turn of the century.

Punctuationalism and Society

In its earlier heyday, Social Darwinism attracted a wide following. Like few other concepts, it opened the door for rationalization of nearly everything. Karl Marx bent it to his purposes as readily as did the captains of industry. While Marx equated evolution with revolution, John D. Rockefeller saw the existing order as the final outcome of a natural process. He preached in Sunday school: "The growth of a large business is merely a survival of the fittest. . . . This is not an evil business. It is merely the working-out of a law of nature and a law of God."[1] Marx, for his part, confronted Darwin with an offer to receive the dedication of *Das Kapital*. Darwin declined.

If we look to biological evolution for analogies that might enlighten our choices in directing cultural evolution, then the punctuational model offers a different course of instruction than the gradualistic model. In fact, many theories of social evolution have focused upon selection among groups (tribes, political systems, or businesses) rather than selection among individuals. They have thus resembled species selection. One interesting punctuational analogy is between the evolutionary stagnation of large, complex species and the resistance of large institutions to change. Without attempting to extract an ethic from nature, we can simply note that neither large, complex systems of government nor enormous corporations readily change their behavior. Like new kinds of species, altogether new kinds of government, including that of the United States, tend to spring rapidly from small beginnings. The same may be said for new kinds of corporations, some of which blossom by virtue of technological discoveries which amount to new adaptations. Marx would have found in the punctuational scheme a logical justification for political revolutions, but then Rockefeller might have rationalized capitalistic victories that issue from revolutionary entrepreneurial innovations. The Marxist analogy of punctuational evolution was, in fact, widely touted by socialists early in the twentieth century, when De Vries and others won a considerable following for their arguments that evolution proceeds by macromutational jumps. Here the socialists were seeking to ground their position in a natural ethic.

Even if we have no particular goal and seek nothing more than heuristic analogies, we cannot help but see parallels between large, sluggishly evolving species and such institutions as the Congress of the United States, straight-jacketed as it is by special interest groups tugging its various members in divergent directions and preventing the whole from moving effectively along any particular legislative course. From all this comes no moral code, religion, or law, but a variety of lessons that we may or may not choose to assimilate.

Finally, I wish to consider a question commonly asked of those who study evolution: are humans still evolving? First of all, one must appreciate that it is an inherent property of cultural evolution that it feeds back into biological evolution. To a shocking degree, we are capable of directing evolutionary change within the human species. To date, the changes we have wrought do not reflect a conscious effort at "improvement"—what is commonly spoken of as "eugenics." Rather, they are largely the products of our efforts to preserve human life, or the consequences of medical care. We retain within our breeding society debilitated individuals who would not easily survive in a cultureless nature. This has been the course of action chosen by Western culture, if not by all human societies. We have also exercised some artificial selection in removing deviant individuals from our gene pool, but our cultural retention of the physically weak has probably been more significant.

On balance, Western civilization has partly released itself from the selection pressures of nature, rather than having imposed more stringent selection. Our species is probably now evolving less rapidly than it was in its early phases. Even so, the punctuational model of evolution implies a less significant impact of our culture on our biological evolution than might be envisioned by the gradualist. The fossil record shows no evidence of human evolution in Western Europe during the entire forty thousand years of our species' existence, and this is no surprise. What rate of evolution can we expect when the lineage that we call *Homo erectus* survived for upwards of a million years without altering enough for us, in retrospect, to change its name? If we restricted our analysis to major, genetically determined adaptive shifts, the punctuational model would imply that our species essentially stopped evolving the moment it was born! Fine tuning has probably taken place, and discrete races have certainly been added, but in Western Europe, no anatomical evolution significant enough to be measured has thus far been documented from the fossil record of our first forty thousand years.

Cultural governance of our biological evolution has the potential to become closely similar to our practice of artificial selection or, possibly, to our genetic engineering of other species. Perhaps we need look no farther than our diverse canine productions to see what we might make of our own species. Do the Chihuahua and the Saint Bernard stand as models of how bizarrely we might mold the human body plan? Quite possibly. There is much evidence that unusual domestic animals arise in at least their incipient form as "sports," or individuals notably different from their ancestors. Human culture has tended not to tolerate "sports" within the human species. Cultural prejudices have been stronger here than simple animal rejection instincts. From our past performance, it seems fair to conclude that no population of humanity that was freakish by present standards would readily be tolerated on Earth.

If major, genetically coded shifts of adaptation are restricted to speciation by way of small populations, then I would venture a prediction. Humans will perhaps evolve significantly in a biological sense only if they escape this planet for parts unknown. If we are to speciate, it will probably be in space, where small, inbreeding populations may someday colonize new worlds. If this suggestion is disconcerting, another possibility is dismaying. Nuclear war may someday depopulate the Earth, leaving small, insular, and perhaps genetically altered populations, one or more of which might nucleate a divergent new race or, possibly, a new species of the human family.

Such speculation is far removed from our immediate problems. Long ago, we may, by choice, have involved ourselves in species selection within our own genus: we may willfully have exterminated Neanderthal. Now, in a more honorable pursuit, we grapple with ourselves to avoid the same fate.

NOTES

Chapter 1

1. Charles Darwin, *On the Origin of Species,* 1st ed. (London: John Murray, 1859), pp. 310-311. (Facsimile of the first edition published by Harvard University Press, 1964.)
2. Theodosius Dobzhansky, *Mankind Evolving: The Evolution of the Human Species* (New Haven: Yale University Press, 1962), p. 181.

Chapter 2

1. Charles Lyell, *Principles of Geology,* 4th ed., vol. 2 (London: John Murray, 1835).
2. James Hutton, *Theory of the Earth, with Proofs and Illustrations* (Edinburgh: Weinheim and Codicote, 1795), p. 200. (Facsimile published by H. R. Engelmann [J. Cramer] and Wheldon and Wesley, 1959.)
3. Charles Darwin, *Journal of Researches into the Natural History and Geology of the Countries Visited During the Voyage of H. M. S. Beagle Round the World Under the Command of Captain Fitz Roy, R. N.,* 2nd ed. (New York: Appleton) p. 292.
4. Ibid., p. 301.
5. Ibid., p. 303.
6. Ibid., p. 310.
7. Ibid., p. 174.
8. Ibid., p. 380.
9. Nora Barlow, ed., "Darwin's Ornithological Notes," *Bulletin of British Museum (Natural History). Historical Series* 2, no. 7 (1963).

Chapter 3

1. Charles Darwin, "Essay of 1844," *Evolution by Natural Selection,* by Charles Darwin and Alfred Russel Wallace, (Cambridge: Cambridge University Press, 1958), p. 111.
2. Nora Barlow, ed., *The Autobiography of Charles Darwin, 1809–1882* (New York: W. W. Norton, 1958), p. 120.
3. Gavin de Beer, M. J. Rowlands, and B. M. Skramovsky, "Darwin's Notebooks on Transmutation of Species," Part VI, *Bulletin of the British Museum of Natural History.* Series, vol. 2., nos. 2-6 and vol. 3, no. 5, 1960-1967, pp. 134-135; see Sandra Herbert, "Darwin, Malthus, and Selection," *Journal of the History of Biology* 4 (1971): 209-217.
4. Charles Darwin, *The Descent of Man, and Selection in Relation to Sex,* 2nd ed. (London: John Murray, 1863).
5. Darwin, "Sketch of 1842," *Evolution by Natural Selection,* by Charles Darwin and Alfred Russel Wallace, (Cambridge: Cambridge University Press, 1958), p. 53.
6. Darwin, "Essay of 1844," pp. 112-113.
7. Gavin de Beer, "Darwin's Notebooks," pp. 134-135.

Notes

Chapter 4

1. Francis Darwin, ed., *More Letters of Charles Darwin; A Record of His Work in a Series of Hitherto Unpublished Letters,* vol. I (New York: D. Appleton, 1903), p. 206.

2. Francis Darwin, ed., *The Life and Letters of Charles Darwin,* vol. II (New York: Basic Books, 1959), p. 12.

3. Charles Lyell, *Geological Evidence of the Antiquity of Man* (London: John Murray, 1868).

4. Francis Darwin, ed., *More Letters,* vol. I, p. 159.

5. Nora Barlow, ed., *The Autobiography of Charles Darwin, 1809–1882* (New York: W. W. Norton, 1958), p. 118.

6. Robert Chambers, *Vestiges of the Natural History of Creation,* 3rd ed. (New York: Wiley & Putnam, 1845).

7. Charles Darwin, *The Variation of Animals and Plants Under Domestication,* 2 vols. (London: John Murray, 1868).

8. Francis Darwin, ed., *Life and Letters,* vol. I, p. 397.

9. Thomas H. Huxley, "The Coming of Age of 'The Origin of Species' [1880]," *in Darwiniana: Essays* (New York: D. Appleton, 1897), pp. 228–229.

10. James Hutton, *Theory of the Earth, with Proofs and Illustrations* (Edinburgh: Weinheim and Codicate), p. 200.

11. Charles Darwin, *The Descent of Man, and Selection in Relation to Sex,* 2nd ed. (London: John Murray, 1875), p. 88.

12. Theodosius Dobzhansky, "Adaptive Changes Induced by Natural Selection in Wild Populations of *Drosophila,*" *Evolution,* 1 (1947): 1–16.

13. Theodosius Dobzhansky, *Genetics and the Origin of Species* (New York: Columbia University Press, 1937).

14. Theodosius Dobzhansky, "Species of *Drosophila,*" *Science,* 177 (1972): 664–669.

15. Julian S. Huxley, *Evolution: The Modern Synthesis* (London: Allen & Unwin, 1942).

16. Julian S. Huxley, *Evolution: The Modern Synthesis* (London: Allen & Unwin, 1942), pp. 32–33.

17. Ibid., p. 389.

18. William Coleman, *Biology in the Nineteenth Century: Problems of Form, Function, and Transformation* (Cambridge: Cambridge University Press, 1977), p. 80.

Chapter 5

1. Loren Eiseley, "Neanderthal Man and the Dawn of Human Paleontology," *Quarterly Review of Biology* 32 (1957): 323–329.

2. Joseph Leidy, "Extinct Mammalian Fauna of Dakota and Nebraska," *Journal of the Academy Natural Science of Philadelphia,* 2nd ser. 7(1869): viii.

3. Ernst Mayr, *Systematics and the Origin of Species,* reprint ed. (Magnolia, Mass.: Peter Smith, 1942).

4. Ernst Mayr, "Change of Genetic Environment and Evolution," in J. S. Huxley, A. C. Hardy, and E. B. Ford, eds., *Evolution as a Process* (London: Allen & Unwin, 1954), pp. 157–180.

5. Ernst Mayr, *Animal Species and Evolution* (Cambridge: Harvard University Press, 1963).

6. Niles Eldredge, "The Allopatric Model and Phylogeny in Paleozoic Invertebrates," *Evolution* 25 (1971): 156–167.

7. Niles Eldredge and Stephen Jay Gould, "Punctuated Equilibria: An Alternative to Phyletic Gradualism," in T. J. M. Schopf, ed., *Models in Paleobiology* (San Francisco: Freeman, Cooper, 1972), pp. 82–115.

8. Claude Delamare-Deboutteville and Lazare Botosanéanu, *Formes Primitive Vivartes* (Paris: Hermann, 1970).

9. Francis Darwin, *More Letters of Charles Darwin; A Record of His Work in a Series of Hitherto Unpublished Letters,* vol. I (New York: D. Appleton, 1903), pp. 20–21.

10. Hugh Falconer, "On the American Fossil Elephant of the Region Bordering the Gulf of Mexico (*E. Columbi* Falc.); With General Observations on Living and Extinct Species," *Natural History Review,* 3 (1863): 43–114.

11. Thomas H. Huxley, "The Anniversary Address," *Proceedings of Geological Society of London,* 1862, pp. xl–liv.

Notes

12. Thomas H. Huxley, "The Coming of Age of 'The Origin of Species' [1880]," in *Darwiniana: Essays* (New York: D. Appleton, 1897), p. 241.

13. Martin J. S. Rudwick, *The Meaning of Fossils* (London: MacDonald, and New York: Elsevier, 1972).

14. Glenn L. Jepsen, George Gaylord Simpson, and Ernest Mayr, *Genetics, Paleontology, and Evolution* (Princeton, N.J.: Princeton University Press, 1949).

15. George Gaylord Simpson, *Tempo and Mode in Evolution* (New York: Columbia University Press, 1944).

16. George Gaylord, *The Major Features of Evolution* (New York: Columbia University Press, 1953).

17. Otto Schindewolf, *Grundfragen der Paläontologie* (Stuttgart: Schweizerbart, 1950).

18. Otto Schindewolf, *Der Zeifaktor in Geologie und Paläontologie* (Stuttgart: Schweizerbart, 1950).

19. Mayr, "Change of Genetic Environment and Evolution," pp. 157-180.

20. Eldredge, "The Allopatric Model," pp. 156-167.

21. Eldredge and Gould, "Punctuated Equilibria," pp. 82-115.

22. Ernst Mayr, *Populations, Species, and Evolution* (Cambridge: Harvard University Press, 1970), p. 183.

23. Martin J. S. Rudwick, *The Meaning of Fossils* (London; MacDonald, and New York: Elsevier, 1972), pp. 265-266.

Chapter 6

1. Theodosius Dobzhansky, *Mankind Evolving: The Evolution of the Human Species* (New Haven: Yale University Press, 1962), p. 183.

2. Theodosius Dobzhansky, "Species of *Drosophila*," *Science* 177 (1972): 664-669.

3. Richard C. Lewontin, *The Genetic Basis of Evolutionary Change* (New York: Columbia University Press, 1974), p. 159.

4. Harlan Lewis, "Catastrophic Selection as a Factor in Speciation," *Evolution* 16 (1962): 262.

5. Charles Darwin, "Essay of 1844," *Evolution by Natural Selection*, by Charles Darwin and Alfred Russell Wallace (Cambridge: Cambridge University Press, 1958), p. 150.

6. Charles Darwin, *On the Origin of Species*, 3rd ed., (London: John Murray, 1861).

Chapter 7

1. Theodosius Dobzhansky, *Mankind Evolving: The Evolution of the Human Species* (New Haven: Yale University Press, 1962), pp. 220-221.

2. Richard E. F. Leakey and Alan C. Walker, "*Australopithecus, Homo erectus* and the Single Species Hypothesis," *Nature* 261 (1976): 572-574.

3. Charles Darwin, *The Descent of Man, and Selection in Relation to Sex*, 1st ed., vol. 2 (London: John Murray, 1871), pp. 389-390.

Chapter 8

1. Theodosius Dobzhansky, *Mankind Evolving: The Evolution of the Human Species* (New Haven: Yale University Press, 1962), p. 6.

2. Carl G. Hempel, *Aspects of Scientific Explanation and Other Essays in the Philosophy of Science* (New York: The Free Press, 1965), p. 370.

3. See Neal C. Gillespie, *Charles Darwin and the Problem of Creation* (Chicago: University of Chicago Press, 1979), p. 14.

4. Charles Darwin, "Essay of 1844," *Evolution by Natural Selection,* by Charles Darwin and Alfred Russel Wallace (Cambridge: Cambridge University Press, 1958), p. 253.

Notes

Chapter 9

1. Julian S. Huxley, *Evolution: The Modern Synthesis* (London: Allen & Unwin, 1942), p. 389.

2. Germaine L. Warmke and R. Tucker Abbott, *Caribbean Seashells* (Naberth, Pa.: Livingston Publishing Company, 1961).

3. Charles Lyell, *Principles of Geology*, 4th ed., vol. 2 (London: John Murray, 1835), p. 391.

4. Sandra Herbert, "Darwin, Malthus, and Selection," *Journal of the History of Biology* 4 (1971): 209–217.

5. Garland Allen, *Life Sciences in the Twentieth Century* (Cambridge: Cambridge University Press, 1978), p. 143.

Chapter 10

1. Quoted in William J. Ghent, *Our Benevolent Feudalism* (New York: Macmillan, 1902), p. 29.

INDEX

aardvarks, 84
Abbott, R. T., 191
accidental evolution, 68–69
Acheulian culture, 159–160
Acongagua (volcano), 19
adaptation: genesis of, 67; secondary sexual, 44–45
adaptive innovation, 94
adaptive radiation: of cichlid fishes, 119; of dryopithecines, 140–141; early Cambrian, 86–90; of flowering plants, 90–91, insular, 126; of mammals, 91, 93–95; of marsupials, 124–126; rapidity of, 122; as site of large-scale evolution, 94; of hydrobioid snails, 123–124
Aegyptopithecus, 140–141
Agassiz, L., 28
agnosticism, 167
algae: blue-green, 195–197; fossil, 7
alleles, 65
Allen, G., 200
Altamira cave, 162
ammonoids, 175–176
amphibians, metamorphosis in, 127–128
Andean mountain chain, 19–20; Uspallata Range of, 21
angiosperms, 90
animal husbandry, 38
Animal Species and Evolution (Mayr), 77
anthropology, 164
antlers, 43
apes, 141
Archaeopteryx, 75–76, 176
archetypal idea of animal groups, 39–40
Ardrey, R., 144
argus pheasant, 45
Argyll, 8th Duke of, 179

Aristotle, 23
armadillo, giant, 28, 30
art, 161–162
artificial selection, 9–10, 41–42
assortative mating, 121
Australia, marsupials of, 124
australopithecines, 142–145; coexistence with *Homo erectus*, 139; pelvic and limb bones of, 145; tools of, 160; tracks of, 145, 150
Australopithecus afarensis, 149, 151, 154
Australopithecus africanus, 142–144, 149
Australopithecus robustus, 144
axolotl, 127–128

Bacon, F., 105
Baer, K. E. von, 37
bairdiellas, 120–121
Bakker, R. T., 91
balance of nature, 52, 197
baleen whale, 39
barnacles: Darwin's studies of, 58–59; species selection of, 187–188
Barombi Mbo (lake), 114
basal neocortex, 158
Bateson, W., 63, 66
bats, 93–94
Beagle, H.M.S., 17, 20, 23–28, 31, 34, 35, 42, 58, 167
beaks: of Galápagos finches, 33; of honeycreepers, 115–116
bears, 98–99, 128–129
beetles, 79
Bering Land Bridge, 104
Big Horn Basin (Wyoming), 95–96

Index

73, 77, 87, 102, 105, 106, 114, 136, 164, 171, 194, 199-200; postponement of publication by, 58-60; on predation, 194; religious convictions of, 17, 167; "Sketch of 1842," 47, 49, 57; and species selection, 199-200; on "sports," strategies of, 54-58; tree of life metaphor used by, 35-37, 114; *The Variation of Animals and Plants under Domestication,* 52, 59, 136-137; view of fossil record, 6-8, 24-26, 40-41, 73, 75, 105-106

Darwinism: and the Modern Synthesis, 67-68; nadir of, 66, 180; social, 203-205;

Davis, D., 128-129

Davis, G. M., 123-124

dawn horse (*Hyracotherium*), 3, 4, 95-97

Death Valley, 113

deduction, 105

Delamare-Debouteville, C., 85

The Descent of Man and Selection in Relation to Sex (Darwin), 46-47, 62, 140

De Vries, H. 66, 205

dinosaurs: domination of, 93; extinction of, 91, 126; mammals in the Age of, 25

directed speciation, 202

dispersal, vectors of, 28

divine creation, 27, 48, 105, 167; forms of, 172, 174

Dobzhansky, T., 14, 67-70, 174; *Genetics and the Origin of Species,* 68; *Mankind Evolving,* 138

dogs, domestic, 38

"domestic productions," 41, 51-52

dominance, allelic, 65

Doyle, A. C., 82

Doyle, J. A., 90

Drosophila, 67-68, 117

dryopithecines, 140

Earth: age of, 19, 61, 108-109; history of, 80-81; as isotopic furnace, 61

ecological opportunity, 94; and quantum speciation, 119

economy of nature, 52

Edentata, 28

Eiseley, L., 72

Eldredge, N., 78, 108

elephants: evolution of, 99-100; fossil, 55, 102

embranchements, 23, 36

embryology, 37-38

emergence, 184-185

empiricism, 105, 171-177

endemism, 31

environment, evolution's tracking of, 11

Eocene Epoch, 95-96

Eohippus (dawn horse), 95-97

epochs, geological, 80

Equus group, 4-5

eras, geological, 80

Erben, H. K., 176

erosion, 20-21; of British Weald Formation, 48

"Essay of 1844" (Darwin), 42, 47, 52, 136, 179

eutrophic lakes, 196-197

evolution, 46; accidental, 68-69; of ammonoids, 175-176; of birds, 75-76; cultural, 161-164; cultural analogs to, 203-207; Cuvier's contributions to, 22-23; of elephants, 99-100; estimation of rates of, 81-82; evidence of, 27-34, 165-177; experimental quality of, 53; gradualistic model of, *see* gradualism; of horses, 4-5, 74-75, 104; human, *see* human evolution; in the laboratory, 67; macro- and micro-, 184; mechanism of, 41-45; and modern genetics, 172; Modern Synthesis of, 67-71, 77, 106-108; parallel, 201-202; pathways of, 73-75, 145; proof of, 169-171; punctuational model of, *see* punctuational model of evolution; quantum, 107; rapid steps of, 57-58, 70, 77-78, 86; reversals of, 151, 155; of snails, 133-134; via speciation, *see* speciation; story of, 177; "straight line," 66-67; teleological view of, 180; testing of, 171; tracking environment, 11; traditional view of, 3; trends in, 182-184; unpredictability of, 181

Evolution (journal of the Society for the Study of Evolution), 107

Evolution: The Modern Synthesis (J. Huxley), 69-70

Index

experimental biology, 63
experiments in paleontology, 82–86
expression of genes, 62
extinction: of dinosaurs, 91, 126; evidence of, 25–27; of giant mammals, 29–30; mass, 22–23; and natural selection, 186; of Neanderthals, 154; of woolly mammoth and rhino, 98

Falconer, H., 55, 102, 105, 106
family, 15; unique chromosomal configuration of, 131
fecundity, 9
finches, Galápagos, 32–34, 51, 165
fish: bairdiellas, 120–121; bowfin, 83–84; cichlid, 111, 114, 117–120; devil's pupfish, 113; fossil, 10; lungfish, 84; snail darter, 114; sturgeon, 84
Fitzroy, R., 17
fixity of species, 41
flowering plants: adaptive radiation of, 90–91; pollination of, 94; species selection of, 188–189
foraminifera, 100
fossil record, 72; of Cambrian Period, 86; Cenozoic, 101; Darwin's view of, 6–8, 24–26, 40–41, 73, 75, 105; evaluations of, 78–83; of flowering plants, 90–91; gaps in, 106; gradualism and, 71; of hominid species, 139–140; of Ice Age, 97–98; Lyell's view of, 24–25, 41; populations too small to leave, 99; of South America, 28–29, 40; and testing of evolution, 171–172, 174–176
fossils, 7; algae, 7; armadillo, 28; brachiopods, 13; Cenozoic, 82; elephant, 55; fish, 10; hominid, 164; ichthyosaur, 24; insects, 78–79; invertebrates, 80; leaves, 12, 91, 92; living, 84–85; mammoths, 25–26; oldest human, 153; pollen, 91; sea urchins, 8–9, 80; snail, 11; Tommotian shelly fauna, 89; trace, 87–88; trilobite, 6, 89; vertebrates, 80
frontal lobes, 158
fruit flies, 67–68, 117

Galápagos finches, 32–34, 51, 115
Galápagos islands, 31–34
Galápagos tortoises, 31–32
Gardens of Eden, evolutionary, 111, 122
Gaudry, A., 75, 103
gene pool, 8–9; complete mixing of, 50; human, 138
genes: alleles of, 65; expression of, 62–65; flow of, 130, 157; mutation of, 65–66; position effects of, 131; recessive, 130; regulatory, 108, 127, 129; structural, 108, 127
Genesis, 172, 174
genetic character, cryptic perpetuation of, 64–65
genetic drift, 12; in small populations, 69
genetics: birth of modern, 62; and gradualism, 107; and Modern Synthesis of Evolution, 67–71; quantitative population, 68, 108; regulatory, 108; and species formation, 126–136
Genetics and the Origin of Species (Dobzhansky), 68
Genetics, Paleontology, and Evolution (Jepsen, Mayr, and Simpson) 106–107
genitalia, of beetles, 79
genus, 15; duration of, 100; endemic, 114; origin of, 98–99, 114; unique chromosomal configurations of, 131
Geoffroy St. Hilaire, E., 22
geographic separation, 121–122
geographic speciation, 49–51
geography of life, 27–34
Geological Evidence of the Antiquity of Man (Lyell), 57
Geological Society of London, 102–103
geology: fossil record of, see fossil record; Lyell's contributions to, 18–19; uniformitarian, 47–48
giant ground sloth, 21, 28, 30
giant panda, 128–129
glacial sediment, 173
glaciers, 97
Goldschmidt, R., 71, 135
Gondwanaland, 124
gorillas, 142
Gould, J., 29
Gould, S. J., 78, 108

Index

Index

molecular biology, 185

mollusks: chronospecies of, 101; dispersal of, 28; extinction of, 192-193

Morgan, T. H., 66

moths: banana feeding, 111-112; peppered, 130

musk ox, 98

mutation, 11, 65-66; chromosomal, 131; detrimental nature of, 67; favorable, 69; of great effect, 66, inadaptive, 107; of regulatory genes, 127

Nabugabo, Lake, 111, 118

natural groupings of animals, 23, 36

natural selection, 9-13; Darwin's conception of, 41-45; Darwin's gradualistic view of, 47-53; evidence for, 27; and extinction, 186; gradualistic view of, 48-49; immunity to, 85; mutation as raw material of, 65-66; paleontologists' view of, 75; and population distribution, 69, 130; and rapid formation of species, 102, 106; reception of, 54-55, 57-58, 60; sexual, 43-45; in small populations, 166-167; and speciation, 184; stabilizing, 121; and transformation of a whole species, 69-70, 130

nature: balance of, 52, 197; mathematics and, 108-109; versus nurture, 62

nautiloids, 175-176

Navaho creation story, 178

Neanderthal, 5, 151-154; basal neocortex of, 158; brain capacity of, 152; culture of, 160; extinction of, 154; ritualistic burial of, 168-169, skull of, 153, 164, throat structure of, 156; tools of, 160-161

neo-Lamarckianism, 76-77

neoteny, 156-158

Noah's flood, 172

Oldovan culture, 159-160

Olduvai gorge, 148-149, 159

Olson, S., 116

On the Origin of Species by Natural Selection, 12-13, 27-28, 30-31, 34-63, 73, 77, 87, 102, 105, 106, 114, 136, 164, 172, 194, 199-200; centennial year of, 70, 77; publication of, 35, 109; reception of, 60-62

optimization, 189

orangutan, 155

Ordovician period, 89

orthogenesis, 66-67, 180-181

Osborn, H. F., 77

Osorno (volcano), 19

Ovibos (musk ox), 98

Owen, R., 39

Paine, R. T., 194

paleontology, 71; decline of, 101-103, 106, 109; hypothesis testing in, 82-86; and the Modern Synthesis, 77, 106-109; and natural selection, 75

Paleozoic Era, 21; life at start of, 89, 102-102

Panama, 125

pandas, convergence of, 128-129

parallel evolution, 201-202

Paris Basin, 21

particulate inheritance, 62

Paul, St., 43

peacock, 43

peas, Mendel's experiments with, 63-65

Peking Man, 146

pelvis, 143; of Neanderthal, 153

peppered moth, 130

periods, geological, 80

"persistent types," 103-104

Peuguenes Ridge, 20

Phacops rana (trilobite), 6

Phanerozoic Eon, 86

pheasant, 45

Phillips, J., 102

phylogeny, *see* tree of life

Pithecanthropus, 146-148

Platonic ideal of species, 39, 41

Pleistocene Epoch, 97-100

plenitude, concept of, 23, 26, 52-53, 189-197

Index